RETURN TO THE RED PLANET

FRONTISPIECE: The four largest volcanoes, the great canyons, and the impact basin of Argyre are prominent on this image of Mars taken June 17, 1976, by *Viking 1 Orbiter* as it approached the planet. (NASA/JPL)

RETURN TO THE RED PLANET

ERIC BURGESS

COLUMBIA UNIVERSITY PRESS NEW YORK

COLUMBIA UNIVERSITY PRESS
New York Oxford

Library of Congress Cataloging-in-Publication Data

Burgess, Eric.
Return to the red planet / Eric Burgess.
p. cm.
Includes bibliographical references.
ISBN 0-231-06942-1 (alk. paper)
1. Mars (Planet)—Exploration. 2. Mars probes. 3. Outer space—
Exploration—United States. I. Title.
QB641.B819 1990
919.9'23'04—dc20
90-1653
CIP

Casebound editions of Columbia University Press books are Smyth-sewn
and printed on permanent and durable acid-free paper

Printed in the United States of America
c 10 9 8 7 6 5 4 3 2 1

In fond memory of my mother, Lily Burgess,
who introduced me to The Princess of Mars in the 1920's

CONTENTS

PREFACE

In 1948 the late Olaf Stapleton, philosopher and author, was invited to address the British Interplanetary Society at the Caxton Hall in Westminster, London on the subject of interplanetary man. To a packed audience of several hundred people Dr. Stapleton stated "It seems very unlikely that there is any intelligent race anywhere in the Solar System, except mankind. Mars is the only planet that should make us hesitate to accept this conclusion, but in this case too the betting is heavily against any highly developed life."

"Would there be any point in colonizing the planets?" asked Stapleton. "What is it all for?" He suggested three motives in addition to scientific curiosity and the adventure of exploration: to exploit the physical resources of other worlds, to increase human power over the environment, and to make the "most" of humans, or the "best" of them. First, however, Stapleton insisted that we try to define a human. He considered the human's uniqueness as being an ability to solve novel problems, the power of abstraction, the ability to concentrate on the relevant and ignore the irrelevant. The major problem, argued Stapleton, was not what we humans are to do with the planets, but what we have to do with ourselves.

He stated that we should use the planets "to afford to every human being the greatest possible opportunity for developing and expressing the distinctively human capacity as an instrument of the spirit, as a center of sensitive and intelligent awareness of the objective universe, as a center of love of all lovely things, and of creative action for the spirit."

"The goal for the Solar System," he said, "would seem to be that it should become an interplanetary community of very diverse worlds each inhabited by its appropriate race of intelligent beings" through which, he contended, "new levels of mental and spiritual development should become possible, levels at present quite inconceivable to man."

He pointed out how mankind was becoming divided between two profoundly different systems of thought and value centered roughly in America and Russia. Each contains important truths which the other ignores. If each side can learn from the other, argued Stapleton, the result could be a far more adult and spiritually enriched humanity than could ever have occurred without the cultural clash of mighty opposites.

We are today, some forty years later, at a stage in human history when each side of these great ideological divisions of humanity is beginning to try to understand the other. When cooperation in many fields is being sought and when a great opportunity is unfolding for humanity to begin a cooperative, international in-depth exploration of the Solar System for the common good.

But as we approach the end of the twentieth century, we need to examine our nation's capabilities and desire to enter a new era of interplanetary man.

In a talk at NASA-Ames Research Center on the occasion of the tenth anniversary of a Pioneer spacecraft in orbit around Venus, Andrew Fraknoi, Executive Officer of the Astronomical Society of the Pacific, pointed out some disturbing facts about science education in American today. He referred to a Stanford University School of Education Survey released in January 1985 which showed that of the twenty-five instructional hours in a school week, the typical elementary school student in the United States does science for only 44 minutes. In the Soviet Union, East Germany, China, and Japan every student spends about three times as much time on mathematics and science as the most science-oriented students in the United States do. Dr. Fraknoi also pointed out that four out of five of our citizens have never had even high school physics.

It is ironical that a society such as ours which has sent men to the Moon and unmanned spacecraft throughout the Solar System should today be ignoring the potential of our youth and denying them the experience of searching for wisdom and understanding, of seeking the universal principles that govern the universe in which they live. This is particularly disturbing in view of the exciting challenge we can offer our youth. In the words of the 1986 Report of the National Commission on Space recomending a bold national plan for the United States: "To lead the exploration and development of the space frontier, advancing science and technology, and enterprise, and building institutions and systems that make accessible vast new resources and support human settlements beyond Earth orbit, from the highlands of the Moon to the plains of Mars."

Mars is, indeed, a beckoning "star" to humanity, a peaceful goal to free us from tribal rivalries. It offers a technical challenge of exploration, a scientific challenge of trying to understand its evolution and its biology, extant or extinct, a political challenge of coming to terms with how it shall be explored either on a national or an international basis, and a philosophical challenge of what shall we do with Mars and where will it lead our species. Most important, it offers a peaceful, worthy, clearly defined, and economically viable goal for humanity.

Just after the first spacecraft landed on Mars and allowed me to look around through electronic eyes on the seeming lifeless rocky surface of the Red Planet, I abandoned the lights of Los Angeles to delight in the dark, star-spangled skies amid the redwoods of

Northern California. When I switch off my computer and kill my office lights and walk the few paces to my nearby house, I often marvel anew at the magnetic brilliance of those planets that wander among the fixed stars of the constellations. Especially I am entranced by Mars for when it is at a close opposition such as that in 1988 it is such a compelling object in the midnight sky. I am reminded of my first published article over fifty years ago in the *Manchester Forum* in 1936, "Mars, Possibilities of Life." "It may not be life as we know it. Strangely different creatures may have received the power of reasoning." I wrote; "But who can say? Life may take on other forms which do not depend upon air or water. Why should man be so conceited as to think himself the sole occupant of the universe?"

Today we do have very different views on life and living things. We see life not necessarily as metabolism but rather as that which sustains a molecular pattern through which flows an enormous number of atoms, and a pattern which has a repair mechanism to preserve its integrity or can replicate to circumvent irreparable damage caused by the inevitability of the changing external environment of a dynamic planet such as Earth. It is unlikely that we will ever find those strange creatures of science fiction elsewhere, but we must be alert to patterns which replicate and which appear to negate the growth of entropy within their environment. These may be the extraterrestrials we are seeking. We need humans on other worlds on a permanent basis to seek such patterns or the evidence that such patterns existed in the past, if we are ever to be sure of whether or not there is other life in our Solar System.

The challenge of other worlds is clearly before us. In this book I have attempted to provide a background on Mars and an outline of some of the steps to Mars that need to be taken during the remaining years of this century. I hope that we will undertake these steps leading to a permanent human presence on Mars early in the twenty-first century. This will be our first definitive move toward adulthood as a planetary species and our first adventure as an interplanetary species.

Eric Burgess
Sebastopol, California

RETURN TO THE RED PLANET

1

ANTHROPOMORPHIC MARS

Mars the mysterious; a baleful red planet that every two years glowers a challenge to mankind amid the constellations of the zodiac, and about every fifteen years becomes one of the brightest objects in the night skies of Earth. As Mars sweeps to within 35 million miles (56.3 million km) of our planet Earth at its closest approach, astronomers observe bright polar caps and what appear to be large-scale seasonal changes in color and surface patterns on the Red Planet.

They see a planet that is somewhat smaller than Earth, rotating on a tilted axis like Earth to give it seasons that are about twice as long as ours, but with a day of about the same length as Earth's. Among all the planets Mars seemed most like our own. But in the second decade after unmanned spacecraft landed on the dusty plains of Mars to ferret among its rocks and to photograph Martian sunsets in pink skies, we still have not answered many mind-troubling questions about the Red Planet as we try to reconcile deep-seated human intuitions with scientific observations and logic. Scientists world-wide are looking forward to a return to the Red Planet.

The fourth planet outward from the Sun, revolving beyond the orbit of Earth, Mars has been and continues to be the subject of major human speculations. The orbit of Mars departs appreciably from a circle. At its closest approach to the Sun (perihelion) Mars is 27 million miles (43.45 million km) closer than at the opposite and most distant point of its orbit (aphelion). An opposition of Mars occurs when Mars and Earth are aligned on the same side of the Sun (figure 1.1). Oppositions occur approximately every two years and two months. If an opposition takes place when Mars is at its closest to

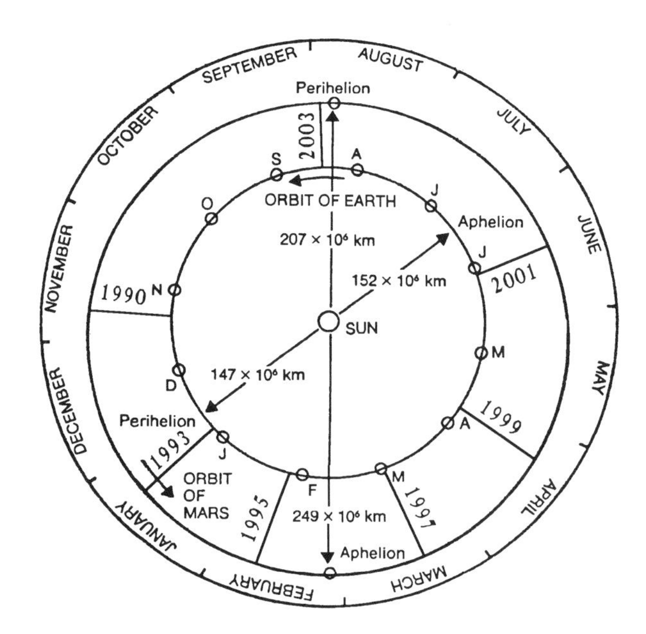

FIGURE 1.1: The orbits of Earth and Mars showing how the distance between the two planets changes greatly at different oppositions. Mars is fourth planet in order of distance from the Sun; mean distance, 141,500,000 miles (227,700,000 km); eccentricity of orbit 0.0933; sidereal period, 687 days; rotation period, 24 hours, 37 minutes, 22.6 seconds; mass, 0.108 Earth masses; diameter, 4200 miles (6760 km); mean density, 3.95 that of water; surface gravity, 0.38g; the planet has two satellites, Deimos and Phobos.

the Sun, the planet is also at its closest to the Earth, about 35 million miles (56.3 million km) distant. At an opposition when Mars is most distant from the Sun, the Red Planet is about 62 million miles (99.8 million km) from Earth. The close oppositions occur on the average every 15 years (eight revolutions of Mars take 15.046 Earth years), but vary between 14.9 and 17.2 years (table 1.1). At such favorable oppositions astronomers get their best views of the planet. Mars has not been of interest only to astronomers. The planet has from time immemorial been of importance in many other ways including most ancient pantheons. Just why the planet Mars should have attained such a lofty status in ancient mythological pantheons remains something of a mystery. In ancient religions Mars had traditionally been designated as the carrier of malefic force, and continually associated with elements of destruction and war. Except for its red color and the association of red with blood, there seems no clearcut reason why Mars should have been accorded such a stature; for although the red color of the planet is pronounced, it is not a brilliant red like that of the setting sun, nor is it blood red, but is closer to pink.

The controversial Immanuel Velikovsky suggested an explanation. He applied a background of history, law, biology, and medicine to amass a fantastic amount of historical data which he claimed challenged the generally accepted views of evolution of the Solar System and its planets. In 1950 Velikovsky upset the scientific world when he claimed in his book *Worlds in Collision* that mankind's views about Mars were a result of turmoil in the Solar System during geologically recent times which resulted in close approaches of Mars to the Earth. He concluded, from his studies of thousands of ancient documents, folk tales, mythologies, and legends from around the world, that Mars collided with other planets and at times rushed toward Earth with its atmosphere

TABLE 1.1. Close Oppositions of Mars

Date	*Distance (million miles)*	*Angular Diameter (arc seconds)*
August 11, 1971	34.74	25.12
July 12, 1986	37	23.3
September 25, 1988	37	23.2
August 28, 2003	34.51	25.3
July 24, 2018	35	24.5
August 15, 2050	34.6	25.23
July 16, 2065	36	23.6
September 30, 2067	37	22.8
September 1, 2082	34	24.9
August 2, 2097	35	24.7

streaming out like a great fiery sword. These close approaches, he wrote, terrified people and led to terror, wars, and bloodshed. When I talked with Velikovsky in the late 1960s he claimed that Nergal, the Babylonian Mars, Ares, the Greek Mars, and the Roman god Mars, were all worshiped as malefic deities because of racial memories of close approaches of Mars to Earth which had caused catastrophes in the heavens and on our planet.

Planetologists cannot reconcile such catastrophic occurrences in the Solar System with what the space program has discovered about the Moon and the planets and what geologists know about the Earth. Major catastrophic events, such as those proposed by Velikovsky, according to the scientific record must have taken place many hundreds of millions, possibly billions, of years ago if they occurred at all.

Another much more recent conjecture that surfaced in science fiction plots and books claiming that extraterrestrials visited the Earth in ancient times, seeks to explain man's preoccupation with Mars by suggesting that Mars may have been a populated world in its distant past; a world with a plentiful atmosphere, with oceans and with highly advanced civilizations from which the first humans came to Earth. There is, however, no evidence to support such theories. Discoveries by spacecraft lend credence to the viewpoint that copious amounts of water once flowed on the Red Planet, but no evidence of life was found at the two *Viking* landing sites nor were ruined cities or other artifacts of life detected from space by the *Viking* orbiters. There are some large unusual features on the planet, of pyramidical and other shapes, which have been suggested as being formed by other than natural processes. However, although these features are fascinating and are discussed in more detail elsewhere in this book, the evidence for claiming them as created by intelligent beings is not very compelling.

Scientists have discovered that liquid water in the form of damp soil or underground pools may still be present at the lowest elevations on Mars today. They base this conclusion on the character of some radar data which implies a 400-mile diameter highly reflective and smooth surface in the Solis Lacus (Lake of the Sun) the so-called Eye of Mars. Generally, however, Mars seems to be a frozen dead world.

Yet it is conceivable to imagine a different state of affairs in the past which ended as Mars began to die as a planet. Global dust storms engulfed the planet and its dust and oceans froze into miles-deep permafrosted regoliths. This ice age has lasted to the present day. Some writers have speculated that as the planet's atmosphere thinned,

Martians desperately fought to survive, scrambling and warring for diminishing economic resources. Memories of this fierce struggle against the planet explain the malefic nature of the planet, claim these highly speculative writings. To escape the doom of their civilization, the Martians sent expeditions to colonize Earth, thereby establishing the human race on our planet. This, the writers add, would account for the abrupt change here on Earth from the developing hominids, who had been around for several million years but had not evolved to civilization, to a new form of man on Earth, descendants of the Martians, which gave rise to Earth's first civilizations. This scenario might explain the sudden appearance of humans here as children of the gods, as stated in many religious creeds, and could also account, so the writers claim, for this unreasonable fear of Mars as a malefic influence and its importance in the pantheon of early civilizations.

There seems no way of proving or disproving such highly speculative theories to everyone's satisfaction. But one fact is paramount. In mythology and in astrology, Mars was and still remains a malefic force, although it can be countered by the goddess Venus. For example, in Renaissance paintings, such as the masterpiece by Botticelli, Mars is depicted as being at rest in the arms of Venus. The Babylonian Mars, Nergal, was a god of mass destruction, of war and of plague. Mars was referred to as "the raging fire-god," "king of battle," and "fire star." The malefic nature of Mars was also apparent in the association of Nergal with the underworld.

Mars at the time of its opposition is a brilliant starlike object in the night sky. Over a period of several months the planet makes a great loop through the constellations. In figure 1.2, Mars loops through the constellation Scorpius, the zodiacal constellation of which, in astrology, Mars is a coruler. The looping effect is caused because the Earth moves along its orbit faster than does Mars and it overtakes Mars when close to opposition. The movements of Mars through the signs of the zodiac were important to many ancient peoples. The Babylonians paid great attention to the planet's risings, settings, and culminations.

The ancient Greeks associated the planet Mars with their god Ares; the *Iliad* refers to Ares (Mars) as a chieftain of valor, a mighty warrior, whose chariot is drawn by flaming steeds. Livy wrote: The Roman people . . . profess that their father and the father of their empire was . . . Mars—the god of war. And the nature of Mars was viewed similarly on the far side of the Asian continent where Chinese astrologers referred to Mars as Ying-Huo, meaning the fire planet. Centuries later and on the opposite hemisphere of our planet, the Aztecs associated Mars with Huitzilopochtli who was described as a destroyer of people and cities, a being of fire. It is remarkable that over such widespread regions of Earth and periods of time Mars should be regarded as an influence of this kind.

In ancient Europe belief in a god of war and battle appears to have come from Greece in the form of Ares, who is described as enjoying nothing but strife, war, and battles. In Greek sculpture this god appears first as a helmeted, heavily armored, bearded warrior, though later he is depicted as a more youthful god. He was honored throughout Greece as the war god accompanied on his campaigns by two squires, Deimos (Fear) and Phobos (Fright).

But by far the strongest mythology of Mars developed among the Romans where he was intimately connected with their history. The story began in early Roman times when the Romans were an agricultural people, and Mars was associated with a god named Silvanus. But as the Romans developed their militarism, Silvanus became

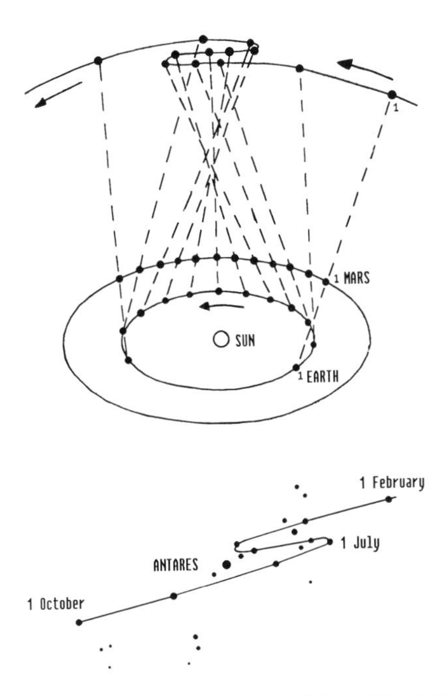

FIGURE 1.2: As Mars passes through oppositions as observed from Earth it traces a great loop in the sky against the background of stars. This is illustrated by the diagrams. a) Shows how Earth overtakes Mars to trace this loop. b) Shows a typical loop through Scorpius, a constellation which ancient astrologers erroneously regarded as being of importance to the influence of Mars on human lives.

separated from Mars who then was accepted solely as a god of war. The name Mars is believed to be derived from Mavors, an older Roman name for the god of war, or possibly from Mamers, which was the name of Mars used by the Oscans, an agricultural people of southern Italy.

The origin of Mars was supernatural in the eyes of the Romans who believed that Mars was the parthenogenetic son of Juno, wife of Jupiter, who was often regarded as a war goddess as well as being the queen of heaven. Mars was not sired by Jupiter but came from Juno's fertilization through mystic union with a flower. This concept probably derives from the earlier concept of Mars as an agricultural god. Married to a relatively unknown goddess, Nerio, Mars commenced an evil career early by raping the beautiful Rhea Silvia, vestal virgin daughter of the King of Alba, surprising her while she slept. Romulus, one of the twin sons who resulted from the forceful union, founded Rome. The colorful legend tells how Amulius tried to drown the newborn children by casting them adrift on the River Tiber, but as they floated down the river in a basket, the river rose and deposited them in front of a temple. There they were found by a she-wolf who suckled the infants until they could be taken care of by a shepherd and his wife. As young men they argued about delineating the city that was to become Rome. Romulus killed his brother Remus and became the founder of the new city. But he could not get women to father his citizens, so, in the tradition of his father, he abducted the daughters of the Sabines to breed his Roman citizens.

Meanwhile Mars himself had not been idle. Despite his marriage and his adventure with Rhea Silvia, he sought to share the bed of the lovely Minerva; but he was outwitted by the goddess Anna Perenna who substituted her own aging body beneath the bridal veils.

Mars had yet another female companion, Bellona, who had mixed relationships with him; as sister, wife, or daughter. She became an important goddess in the Mars cult served by priests picked from the best gladiators. It was, indeed, at the temple of Bellona that the Roman Senate received foreign ambassadors. There, too, the Romans declared war when one of Bellona's priests struck his lance against the war column in front of the temple. Mars even replaced the name of Ares in Roman versions of Greek legends, but the Greek symbol of a bearded, armored, and helmeted war god was adopted as the image of the Roman god. The squires of the Greek Ares were also adopted by the Romans. These became the horses Deimos and Phobos, which pulled the chariot of Mars. Appropriately named for steeds of the god of war, the thundering hooves of Fear and Fright heralded the onslaught of mighty Mars. Before a Roman warrior went off to battle he invariably made sacrifices to Mars, and after each victory the god of war was awarded a share of the spoils. Mars supported his warriors by appearing on the field of battle accompanied by other gods and goddesses; Bellona and Vacuna, Pavor and Pallor, and Honos and Virtus.

Despite his passions and social shortcomings Mars, as father of the founder of the city which became the hub of such a mighty empire, was venerated by the Romans. The militaristic imperialists of the empire needed the spiritual backing of such a god of military strength who could strike terror to the hearts of their enemies. Small wonder that, as the boundaries of the empire spread wide across the ancient world, the Mars cult attracted so many adherents that there were periods when the worship of Mars became stronger than that of Jupiter in influencing the Romans and their culture.

During the early period of the Roman state, when Mars was associated with agriculture, he received the title Gradivus (from the Latin *vern grandiri* meaning to grow large). When he evolved into the god of war for the expansion of the Roman empire, he retained the title Gradivus but its meaning was accepted as being associated with marching troops (*gradi* in Latin meant to march).

The reason that Mars has held such importance in the minds of men for centuries through the present day probably derives from the fact that Mars permeated mythology of the ancient Mediterranean world, a mythology which has colored so much of Western culture and thought. This is especially so regarding the mythology of the Romans whose empire and thoughts spread far and wide over the European continent and spilled beyond its boundaries into Asia and Africa. While probably unaware of the pervading influence of Roman mythology on his thinking, the Victorian who believed so strongly about life on Mars had accepted many of these ancient beliefs into the background of his consciousness. Mars possessed an ancient reality all of its own; first as a god, then as a planet, and later as an abode of extraterrestrial beings.

The strength of the ancient mythology of Mars received somewhat of a setback for a time when it was challenged by the introduction of the Christian era. But it did continue, although remaining essentially underground, until the Copernican revolution broke many of the barriers imposed by centuries of theological dogma to the point that people could again talk openly about the mythology of the planets and even about life being upon them without fear of burning at the stake or facing the dreaded Inquisition formed to purge Europe of Arab and Islamic thought.

With the Renaissance the mythologies resurfaced. The idea of Mars as a god of war survived from then almost into modern times, primarily because of the tremendous influence of classical learning during the Renaissance. Sculpture, art, writing, poetry, all drew heavily on classical concepts and ideas, and the mythological figures became firmly reestablished in human thought. This was a formative period of human history during which, from the seventeenth to the mid-eighteenth centuries, interminable debates raged among scholars about the place of classical mythology in modern literature. Many translations of the writings of Homer, Ovid, Livy, and other great scholars of the ancient world were published, and subtle changes were introduced into the old mythologies. While the religious beliefs of the ancient legends were played down, the morals of the mythological stories were emphasized. Mars survived as one of the important pagan gods. In tapestries, paintings, and sculptures Mars is depicted as an armored man, often with a whip in his hand and accompanied by a wolf—an animal sacred to Mars in Roman mythology. His Roman warlike nature was modified somewhat so that he represented morally good qualities of the positive and the masculine, while retaining the passions of the Roman god.

But his importance did not wane. Even today he is associated with the third month, March, and the third day of the week, Tuesday; both of which in the Latin languages are etymologically traced to Mars. While January was derived from Janus, the sun-god, and February from *februa,* another important Roman festival of expiation, the month associated with Mars, March, was the beginning of the Roman year. The word for March in French is *mars;* in Italian it is *marzo;* in Spanish it is *marzo;* and in German it is *Marz.*

After Sun-day and Moon-day, Mars-day is the next day of the week. The word for Tuesday is intimately connected with Mars in the languages derived from Latin: *mardi* in French, *martedi* in Italian, *martes* in Spanish; and even in English, Tuesday derives from the old English *Tiw,* who was a god of war, akin to the Scandinavian *Tyr,* and the *Ziu* of Southern Germany.

This retention of Mars in third position in the months of the year and the days of the week is therefore a pointer to how important the mythological Mars has been in human thought.

Mars has been important to astrologers. That the birth of a child at a particular time and place governs its life has long been a popular belief. Astrology claims to define the influence that the Sun, Moon, and planets have on each of us relative to the particulars of our births. The fact that man cannot control but can observe the positions of the stars and the planets for a given place and a given time provided a powerful argument for relating cosmic configurations to human lives, linking each to some purpose or plan beyond ourselves. This need to believe in some cosmic plan is deeply seated in human consciousness and surfaces in many ways.

Today Mars has very much the same astrological characteristic as during the Renaissance—a malefic force, but a lesser one than Saturn. And this astrological concept of Mars goes back to ancient civilizations. Astrology was originally identified with astronomy and was based on the assumption that a careful study of the stars and the planets could serve as a guide through life. Astrology played an unfortunate role in history for upon the interpretations of astrologers many rulers, even today, base their actions. All Chinese astronomy, for example, was directed toward observing the heavens for astrological purposes. Their general viewpoint was that each planet moved at its own discretion and by these movements affected the lives of people on Earth. Chinese records from about 400 B.C. report an incident when Mars was close to the three bright stars of the zodiacal constellation Scorpio that include Antares, also a red star. The astrologers warned their ruler of this malefic configuration. Even today the astrological

Scorpio (which now differs from the astronomical constellation Scorpius) is regarded by astrologers as being significant to Mars, showing the persistence and deep-rooted nature of such ancient ideas.

The astrologer's task is to draw a figure of the heavens divided into twelve houses, or equal divisions of the sky, six of which are above the horizon and six below. On it he marks the zodiacal signs and the planets in their proper places. If this is done for the birth of a child it provides the natal chart, and the child's life is supposed to be guided by these configurations. At other times such a chart is a horoscope (derived from the Greek meaning a scrutiny of the hours) to guide a person at a specific time. This is the type of astrological guide met with in the popular press of today.

While today, astronomy and natural science in general discredit astrology, its role in human history cannot be forgotten, nor can its current erroneous influence on many people be ignored. During the many centuries of scholasticism, astrology did not play a major role because logic and faith were the primary ways of thinking favored by the Christian church in Europe. But there were great changes ahead as scholasticism's hold on human thought was challenged by new ideas. The German theologian and cardinal Nicholaus of Cusa (1401–1464), who helped to bring an end to scholasticism, blended scholastic theology with mystic pantheism and predated Copernicus in accepting that the Earth revolved around the Sun. The humanists advocated a new learning that spread through Europe after the fall of Constantinople in 1453. This movement, while not anti-Christian, tended to paganism as it revolted against logic and faith as absolutes and against other religious limitations on knowledge, and revived classical learning.

A tendency to fuse the astronomical, astrological, and mythical aspects of the planets probably stemmed from this early Renaissance period, before astrology separated from natural science in general. At that time the science of the influence of the stars had been rated highly and played a major role in determining the lives of individuals and states. This astrological learning began to occupy more and more human thought and, to the consternation of the church, it started to overshadow much of the sacred learning. The astrological concept of causality from the stars began to usurp causality from God in the human mind. This belief in events beyond the control of man dominating human life even penetrated natural philosophy, to affect concepts of physics and other natural sciences then developing rapidly. There was a firm human belief that an intimate and unbreakable bond existed between the infinitely great and the infinitely small. The intimate bond between the microcosm and the macrocosm also remained basic to the practice of medicine. To account for the origin of syphilis, for example, physicians at one time invoked the baleful conjunction of the two malefic influences, Saturn and Mars.

During the early days of natural science the savants of our world saw no dichotomy between astrology and natural science; the dominion of the heavenly bodies over all things happening on Earth was taken for granted as the natural order. This belief did, however, deny any free will to humans, a denial which became its downfall in the eyes of natural scientists who were seeking ways to understand and control nature instead of being governed by nature.

As humanism spread, thinkers began to reject the idea that the Earth was subjected to the heavenly bodies and that humans were mere puppets pulled by invisible strings from the stars and the planets. In this intellectual climate, the Swiss physician and naturalist Paracelsus (1493–1541) emphasized the importance of direct observations of nature. He stated that man is more than Mars and the other planets, that life is chemical, and that man must look to chemistry for how his life is governed, not to the stars. The idea that mind was superior to matter also began to take hold, but the deep seated primitive human belief of the influence of the planets and the stars still lurked in human consciousness and it survives even today, as witnessed by the popularity of astrology in magazines and the daily press.

The two houses of Mars are Aries, of which the planet is a prominent ruler, and Scorpio, of which it is a coruler. The symbol or glyph for Mars as used in astrology, and also in astronomy as shorthand for the name of the planet, consists of a circle (representing spirit) surmounted by an arrow (representing matter). Other interpretations of the symbol is that it represents the spear and shield of the god of war. This symbol is also used to represent the masculine. In astrology the arrow of matter above the circle of spirit indicates the Martian tendency for the desires of the flesh to surmount the aspirations of the spirit. The accent on sensuality today, as evidenced by questions in letters to the columns of popular magazines and newspapers and the deluge of how-to books, implies that modern men and women are still governed strongly by the Mars of mythology and astrology. But historically, astrology illustrates a relationship between man and the planets that is every bit as valid and as real as the relationships between man and the planets that are illustrated by mythology and religious beliefs, and the progressively more refined and relatively young natural observational sciences of astronomy, physics, and biology.

According to astrology, a child born under the influence of Mars will have a desire to command and influence others, to have power over them in a subjection of others to the self, a way of life that still has many adherents today. While a child of Venus will have a pacifist nature, the child of Mars is the warmonger, and represents the masculine urge to win and succeed, to strive and achieve one's desires. The influence of Mars also represents the animal in man, his basest emotional desires and aggressions.

The invention of powerful telescopes widened the dichotomy between the pseudoscience of astrology and the true science of astronomy. As the resolving power of these instruments increased, scientists were able to see clearly that the celestial objects that are so basic to the art of astrology were, in fact, worlds in their own right. They are places on which people might wander and look out from them, sometimes to see our own Earth as itself a "wandering star." And the advent of the space age generated even greater force fields of barriers between the mystic art and the logical science. What would happen astrologically, for example, to a terrestrial born on a colony on Mars? How could astrologers relate the configurations of the Sun, Moon, and planets as seen in Earth's skies to those in the Martian skies? Science was forced to reject astrology as completely illogical.

Both the religious worship of the planets that was expressed in ancient mythologies of gods and goddesses, and the mystical beliefs of astrology that placed the planets in a direct causal relationship with the destinies of individual people, were profoundly affected by our ever-increasing scientific knowledge of the Solar System. But this knowledge itself, while supplying more technical information, had the effect of opening the planets to man's imagination. In turn this burgeoning knowledge spawned a new attachment with Mars which expressed itself in a new mythology, the literature and films of science fiction.

Modern mythology of Mars began at the beginning of the twentieth century with the novels of Jules Verne and H. G. Wells, followed later by those of Edgar Rice Burroughs,

Ray Bradbury, and others, and a veritable deluge of short stories by the many science and fantasy fiction writers of the pulp magazines published during the 1920s and 1930s.

This fiction emphasized two significant thoughts. The first was a modern expression of the old belief that intelligent beings exist elsewhere in the universe and that many of these beings are far advanced over we terrestrials. The second thought was an old belief also, that some of these beings are human enough to want to control and manipulate terrestrials.

Natural science was supporting the first thought, suggesting that life might be possible on Mars within our own Solar System, and that the laws of probability suggested that there must be millions of other planetary systems throughout the Galaxy. There was no scientific evidence to support the second thought except the example of our own behavior throughout history toward other species of our planet.

But the science fiction writers of the new mythology very often assumed that extraterrestrials would most often be creatures that would want to dominate life on Earth; and this theme still continues today despite the exploration of many planets of the Solar System without discovering life forms there.

Mars figured prominently in the new mythology. It was the Red Planet, the home of the Martians, who were a danger to mankind as they cast envious eyes on a water-rich, plentiful Earth. The science fiction of Mars is fascinating in that it is strongly slanted toward the old notion of aggression and war, of fear and panic. Martians invade Earth or are about to do so. And echoing Kepler's question quoted in *The Anatomy of Melancholy,* "Are we or they Lords of the World?", Martians are often pictured as far advanced intellectually and in science and technology, and with godlike capabilities to harm Earth people by invading our planet or killing our space explorers when they visit Mars.

The nature of the new mythology of Mars was thus one of applying the old mythological Mars into the world of high technology. H. G. Wells in his *War of the Worlds* (1898) depicted the Martians as minds controlling huge mechanical bodies that could move "with the speed of an express train" and shoot destructive beams of intense heat. He described the rolling wave of fear that engulfed London as the great war machines, impervious to barrages of explosive shells, converged on the British capital.

The deep-rooted belief in life on Mars resurfaced like an underwater-launched missile in 1938 when Orson Welles adapted H. G. Wells' *War of the Worlds* to a Halloween radio broadcast. Thousands of people accepted the fiction as fact and fled from their homes. They accepted the mythology even though the time-scale on the newscasts was completely ridiculous; e.g., the flashes of the invaders leaving Mars were reported on the same "newscast" as their arrival on Earth near Princeton, New Jersey, and the subsequent deployment of the Martian war machines and destruction of defensive forces.

Yet recent interviews with other nontechnically oriented people who heard the broadcast showed that many others accepted it as entirely fictional and were not in the least perturbed by the idea of a Martian invasion, even in the credible atmosphere created by swift Nazi occupations of European countries. "We were going to a Halloween dance, and we went," said one woman who was a teenager at the time. "We didn't take the broadcast seriously."

In 1911, when he was in his mid-thirties, Edgar Rice Burroughs wrote an imaginative story about a Captain John Carter visiting Mars through teletransportation. Under the pen name of Norman Bean the fiction was published in six parts as a serial in the magazine *All Story*. Originally entitled "Under the Moons of Mars," the romance was later (1917) published as a book entitled *A Princess of Mars*. Later novels which recounted the adventures of Carter on the Red Planet used the Tarzan formula with great effect to weave stories about a quite fictionalized and unreal Mars which provided a vehicle for fantasies of strange beasts and even stranger intelligent beings, all emphasizing the warlike and horrific nature of the Red Planet. Some biographers claim that Burroughs was influenced by even earlier stories about science fiction heroes visiting Mars; books such as Percy Greg's *Across the Zodiac* (1880), Robert Cromie's *A Plunge into Space* (1891), Ellsworth Douglass' *Pharaoh's Broker* (1899), and Edwin L. Arnold's *Lieutenant Gulliver Jones* (1905).

While the pulp magazines such as *Amazing Stories* featured many stories about Mars, the next major novel was by Ray Bradbury whose *Martian Chronicles* was published in 1950. Bradbury used a purely fictional Mars, in no way related to the real Mars as scientists understood the planet at that time, to show the failings of man. Arthur C. Clarke, in his *Sands of Mars* (1951), tried to show more realistically what mankind might do with the Red Planet from a technological standpoint.

These and many other fictional works, often based upon an increased understanding of our Solar System, created alien worlds and gave them moral philosophies, the trappings and traditions of life, a technological background, and an identity. Mars was no exception; and many engineers and scientists in the space program would admit that they had been intrigued in their adolescent years by the Barsoon of Edgar Rice Burroughs, or the Martians of Ray Bradbury, or the colonization of Mars as visualized by Arthur Clarke.

But even more generally, man as a species seems to have an archtypal need to see himself reflected in the universe, and he uses Earth-bound conventions for his purposes in molding the shape of alien civilizations. While holding on to a great desire not to be alone in the universe, he also fears that any other forms of life will be inimical to him. Mars as the most likely abode of life elsewhere in the Solar System played a very important role in satisfying this human need and focusing this human fear.

As well as written fiction about Mars, the modern Martian mythology has also been expressed in well over fifty motion pictures produced with the Red Planet as their venue, none of which projected any deep feeling of reality about Mars and most of which had trite stories. Beginning with Thomas A. Edison's *A Trip to Mars* (1910), there were many poorly produced films of Martian adventures during the silent era. Flash Gordon was the hero in a *Trip To Mars* (1938), and there were radio series, too. In 1950 the low-budget *Rocketship XM,* pushed for release ahead of George Pal's epic *Destination Moon,* carried a crew (including one woman) to Mars where in a sequence of pink-dyed film the explorers found savage, mutant hordes that had survived a nuclear war. The malefic nature of Mars was very evident; astronauts were captured by the mutants and the hero (Lloyd Bridges) and his girlfriend died during uncontrolled reentry of the spacecraft into Earth's atmosphere.

George Pal produced *War of the Worlds* in 1953, a colorful motion picture based on Wells' book and replete with excellent special effects. But, of course, all the action took place on Earth, as it did in the original novel, and the motion picture depicts only the Martians and their high technology of war machines, not the planet. Pal's contribution to depicting Mars in motion pictures came with his *Mars Project* (1955) which was based on the book by Wernher von Braun, *Mars Project* (1953). Unfortunately the dramatization was poor, most of the action took place on an Earth-orbiting satellite, and Mars was depicted as a very uninteresting and dull world.

All the other motion pictures having Mars as a theme were of little importance or of lasting interest. Plots were often ridiculous and special effects very unconvincing. The most sophisticated of the films of Mars appeared in the late 1970s when Bradbury's *The Martian Chronicles* was aired as a very well-produced, three-part television special. But it was disappointing because the original stories were so out of touch with the real Mars that *Chronicles* could be regarded only as fantasy. Even so it reiterated the theme of a malefic Mars.

As the twentieth century moved into its second half, a new age of mythology emerged, no doubt encouraged by the plethora of science fiction, and expressed in part through beliefs in a modern form of astrology, paranormal psychology, psychic phenomena, demonism, mysticism, the healing power of masochistic exercise, and UFOlogy. Fantasy fiction created new myths of creatures and worlds beyond objective reality immune from the laws established by science as being universally valid. And many people started to accept the myths as realities.

People again started to accept the reality of supernatural beings, of demons and evil powers, of "little green men" in flying saucers, of aliens visiting Earth, of galactic empires. Cults developed that accepted as fact speculations that "gods" had once visited Earth and were preparing to come again. In many ways man's emotional, imaginative, and intellectual association with Mars was integrating ancient mythology with the new mythology of science fiction.

Concurrently, a new era of pseudoscience, as involved and complex as the ancient pantheons and their myths, generated thousands of devotees ready to believe in anything that might help them escape the awesome realities of instabilities of nations and economies, of collapsing traditional values and social structures. And pseudoscience erupted throughout society; politicians and scientists, as well as people in general, extrapolated current events into future trends and accepted the results as real instead of as mere probabilities. Powerful digital computers aided the purveyors of pseudoscience who claimed that their computer models showed what had been and what was to come. Major political decisions affecting economically the lives of billions of people were based on these new myths as enthusiastically and as absolutely as ancient statesmen based their decisions on the configurations of the stars and the imagined pleasures of their gods.

Although literature had had no difficulty in peopling alien worlds with many life forms, the telescopic exploration of the Solar System had proved very disappointing. No reasonable evidence had been found of life elsewhere than on Earth. Yet there was still hope that life would be discovered on Mars.

Against a background in which individuals were increasingly relegated to the role of pawns in some cosmic game between super- or sub-human beings, a scattering of intellectuals offered optimism for a new breed of mankind. Man was becoming interplanetary man in his thinking, but the pace of reality was much slower than the pace of the visionaries. The search for life elsewhere was proving to be a long, tedious, and expensive task. But optimism remained that there would be life on Mars. The optimists visualized an evolution of at least a part of mankind into interplanetary man who would build settlements in space and establish colonies on other worlds. These new people would be free at last from the primitive traditions and beliefs, and the physical and mental limitations of a single planet. High on their target list was a plan to establish a colony on Mars.

The optimism derived from a group of scientists and engineers, including planetologists and biologists, who decided to explore Mars to find out if life does indeed exist anywhere else in the Solar System. A plan had been developed in the 1940s. Early in the space age it was still fragmented among individuals working separately to devise a machine that could reach and land on Mars. Eventually it became part of a long-range program of the National Aeronautics and Space Administration: first, fly spacecraft past Mars and obtain pictures of its surface; then, orbit the planet to obtain detailed maps; finally, land a chemical laboratory on the Red Planet and sample its soil and rocks, searching for living things.

Armed with a completely new view of the Red Planet from orbiting and lander spacecraft, the optimists would no longer have to fear any of the old and traditional aspects of the Red Planet. They no longer had to accept it as any kind of malefic force but as an unintelligent mass of matter waiting to be molded to their will. They accepted Mars as a new economic commons for mankind, offering a challenge to their technology and an opportunity to create a new Eden by refurbishing the Red Planet as a smaller version of Earth. These modern optimists were about to see the Mars of ancient and modern mythology vanquished by twentieth-century science and technology. The dreams were about to become realities as a new Mars of benevolent promise emerged from the womb of the human mind.

First, it is important to recapitulate on the background derived from telescopic observations of Mars made from the Earth.

Intriguing mysteries of Mars as a planet have stirred popular imagination since the building of powerful telescopes, and they reached a first climax toward the end of the nineteenth century. That period was a fruitful age of discovery in natural science and of applications of these discoveries. If, during the 1890s, a well-read man had been asked if he believed there were Martians, he most probably would have given an affirmative answer. His imagination had been piqued by a bewildering array of discoveries in science and by new technologies. Darwin and Wallace had outlined the theory of evolution to account for the diversified life forms of the Earth. Bell had invented the telephone, and Hertz had demonstrated the seemingly miraculous wireless transmission and reception of electromagnetic waves. Railroads were spanning continents, and Langley had flown his steam-powered heavier-than-air *Aerodrome Five* across the Potomac. There were typewriters, electric motors, an electric light, steam tricycles and horseless carriages.

A philosophy of the unity of Nature had been firmly established by discoveries in natural science. Wehler had demonstrated that there was no fundamental difference between so-called organic and inorganic materials; they all consisted of the same basic elements. Moreover, spectroscopic analysis of the Sun and stars had shown that they, too, were made of the same elements as those of our Earth. People accepted that if living things on Earth had sprung up from the nonliving elements, the same processes could have occurred on other worlds. So why should there not be Martians?

It was natural for such a nineteenth-century man to give free reins to his own imagination about Mars. He would probably have said that Martians were far advanced over the civilizations of Earth because Mars was then thought to be an older planet. It was further from the Sun and, according to contemporary theories about the origin of the Solar System, Mars must have been formed before the Earth.

We might wonder why a man of the Victorian age would have had such strong views about a planet that appeared only as a shimmering hazy red disk with indistinct blue-green markings when seen through most telescopes of his time.

The idea of an inhabited Mars goes back almost to the time of the invention of the telescope. In 1659, Christian Huyghens made the first drawing of Mars. It showed a prominent triangular shaped surface feature which is now called Syrtis Major. In his book *Cosmotheoros* he speculated about inhabitants of the Red Planet and their characteristics. But a half century went by before Mars was seriously studied during a set of favorable oppositions in the late 1700s.

At the favorable opposition of 1779, William Herschel used an improved telescope to observe the Martian polar caps (which he attributed to ice and snow) and color changes on the planet. He also established when the seasons begin on the Red Planet, and stated that the inhabitants of Mars probably enjoyed conditions similar to those on Earth because Mars has seasons, and days and nights similar to those of Earth. Johann Hieronymus Schroeter observed the planet between 1785 and 1802, and coined the term aerography (from Ares, the Greek god of war) for use in describing the surface of Mars as geography is used for the Earth. Following the favorable opposition of 1830, Wilhelm Beer and Johann Heinrich von Madler published the first definitive map of Mars in 1832, and Richard A. Proctor published a map in 1834 on which he named features after those astronomers who had been active in observing the planet. Other astronomers who made significant observations of Mars before the 1877 opposition included Angelo Secchi, William R. Dawes, Friedrich Kaiser, and Joseph N. Lockyer.

Popular excitement about Mars as a planet was enhanced by two discoveries made during the very favorable opposition of August 1877 when Mars was close to its perihelion and only 34.8 million miles from Earth. Mars seemed a rather odd planet in that despite searches by astronomers it did not appear to have a satellite. Earth has its Moon, and Jupiter, beyond Mars, possesses a family of satellites. At the favorable opposition of 1830 Madler had sought Martian satellites. He was unsuccessful. Heinrich L. D'Arrest also failed in 1862, and the poet Tennyson referred to the Red Planet as moonless Mars. But at the 1877 opposition Asaph Hall tried with the 26-inch Alvin Clarke refractor of the Naval Observatory, Washington, D.C., and discovered two small satellites of Mars, which he named Deimos and Phobos. These proved to be unusual satellites in that they are extremely small bodies, more like large mountains in orbit. They are thought to be captured asteroids which travel in orbits very close to Mars, so close that they are difficult to see against the glare from the planet.

The strangeness of Hall's discovery was that the presence of these miniature moons had been anticipated several times in a kind of racial *déjà vu.* Johann Kepler, who relied upon observations of the motion of Mars relative to the background of stars to formulate his famous laws of planetary motions, is reported to have stated that Mars might have two moons. And François Marie Arouet de Voltaire in his novel *Micromegas,* published in 1750, wrote about two moons of Mars that had escaped the gaze of Earth's astronomers.

But even stranger was the reference in 1720 to the moons of Mars in Dean Swift's *Gulliver's Travels.* The Laputan astronomers in Swift's novel spoke of two lesser stars, or satellites, of Mars revolving close to the planet in orbits very similar to those of the two satellites discovered 175 years later by Hall. Perhaps Swift had access to Kepler's work, and Voltaire took note of Swift's fiction.

Mars' two satellites are important to the continued exploration of the planet. They

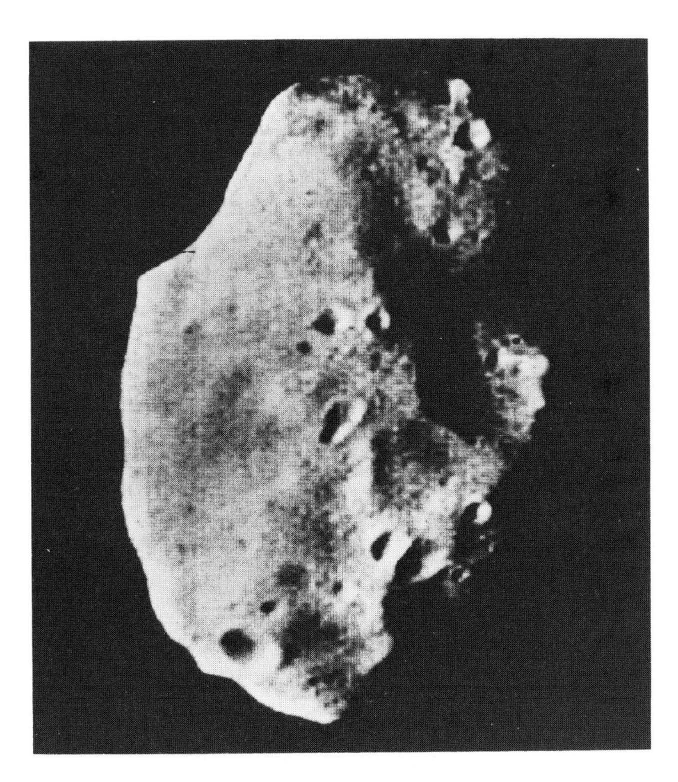

FIGURE 1.3: A computer-enhanced image of the larger Martian satellite, Phobos, taken by *Mariner 9*. The picture shows the irregular cratered surface of the satellite which is possibly an asteroid captured by Mars. (NASA/JPL)

circle the planet close to the equator and are natural space stations for Mars. What might be accomplished on a terrestrial space station can also be accomplished on the Martian satellites.

Phobos (figure 1.3), the larger satellite, orbits Mars in 7.65 hours at an altitude of 3718 miles (5984 km) above the surface. When observed by an astronaut on the surface of Mars, Phobos would rise in the west and five hours later set in the east. Deimos (figure 1.4), the smaller satellite is 12,400 miles (20,000 km) above Mars. It orbits in 30.3 hours so that it rises and sets more normally in east and west. Both satellites are dark colored and have cratered surfaces. They are believed to be similar to carbonaceous asteroids and could be a source of materials for rocket propellants. If so they would make ideal staging stations and considerably reduce the mass needed in Earth orbit for missions to Mars. This will be extremely important for manned exploration of the Red Planet.

The second surprise about Mars was when Giovanni V. Schiaparelli, an astronomer working at Milan Observatory, announced he had observed fine straight lines stretching for many hundreds of miles across the surface of Mars (figure 1.5a). This discovery caused a fierce debate which continued for nearly a century; in fact, until spacecraft flew by Mars and photographed the Red Planet.

Other astronomers who confirmed Schiaparelli's observations included Camille Flammarion in France. Subsequently many books were published claiming that Mars was the abode of life, and could very well be a second but smaller Earth. But there were many equally astute observers who said they could not see the elusive fine lines. Linear

FIGURE 1.4: The smaller satellite, Deimos, is also irregularly shaped with its surface pockmarked by craters. It, too, is believed to be a captured asteroid. Deimos is the outermost of the two satellites. (NASA/JPL)

markings were included on maps of Mars before Schiaparelli drew attention to them. They appear on drawings by Lockyer, Proctor, Dawes, Secchi, and other experienced observers and on the old maps of Mars (figure 1.5b), but none of the early observers had attached significance to the lines.

Following Schiaparelli's announcement, other astronomers at subsequent oppositions gradually began to include the fine lines in their drawings of the planet. In 1888 Henri Perrotin reported seeing the lines with the 30-inch telescope at Nice Observatory. William H. Pickering claimed he saw them at the Harvard Station in Arequipa, Peru. Pickering also claimed that they appeared in the dark areas of Mars as well as the reddish-ocher desert regions. At first very hesitantly and later more freely, other astronomers began to include a few lines in their drawings, and then a veritable canal fever developed. Astronomers in many countries drew maps of Mars that were interlaced with the fine lines of the canals, and they continued to do so for three-quarters of a century. Even at the beginning of the space age in 1957, maps of Mars were still being published that showed many canals. One of the big arguments for intelligent beings being responsible for the lines was the doubling of some of them at certain Martian seasons, first observed in 1881 by Schiaparelli. While single lines might be accounted for by natural processes, the opponents of the speculations about Martian canal builders found it hard to explain how some of the canals could double as a result of natural processes. But again, was the doubling only an optical illusion?

Before 1877 the planet Mars had not been the focus of significantly more attention by astronomers than had the other planets, but Schiaparelli's discovery changed every-

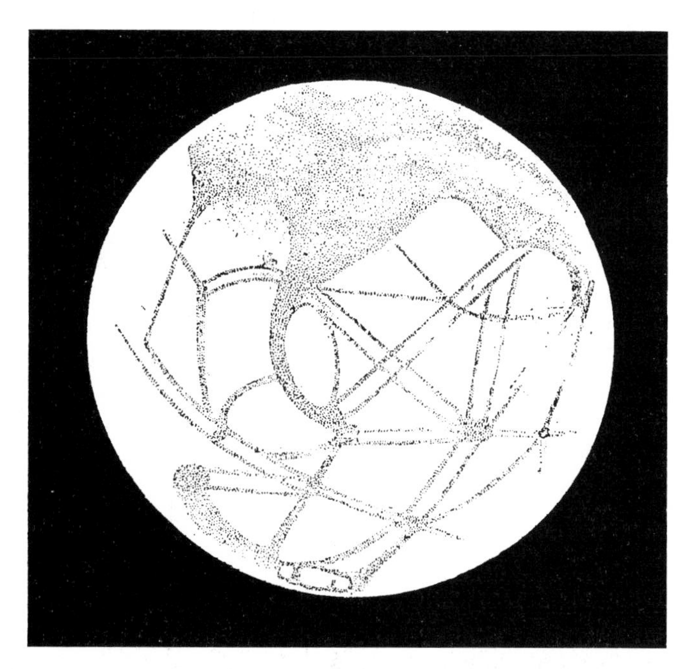

FIGURE 1.5(a): Schiaparelli mapped Mars through the clear skies of Milan during the favorable opposition of 1877. He mapped many fine lines crossing the light areas, features that he called "canali," meaning channels. These channels were loosely translated into English as canals, and this gave rise to a great controversy that raged for almost a century about the reality of these Martian features and whether or not they indicated that the planet was inhabited by an advanced race of intelligent beings. This drawing of part of Mars shows the canals as Schiaparelli claimed to have seen them.

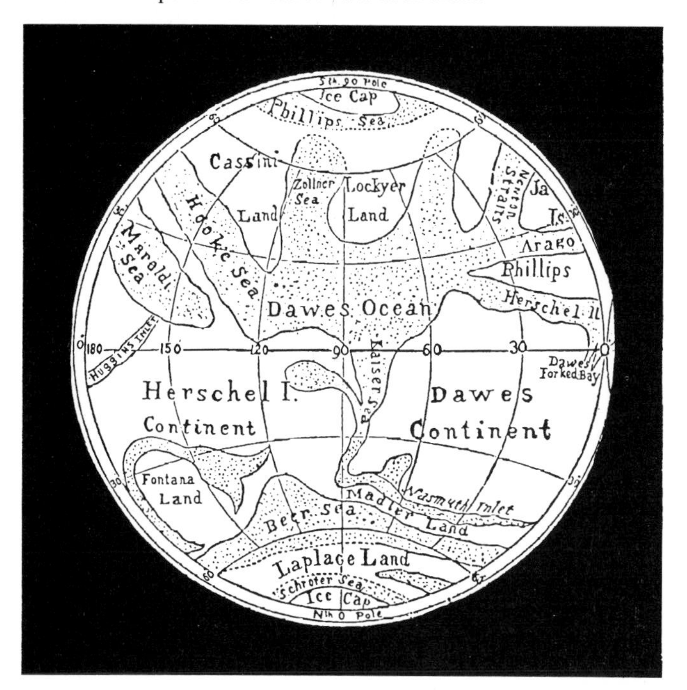

FIGURE 1.5(b): Maps of Mars by Richard A. Proctor (1867) showed several lineaments in positions later identified by Lowell and Schiaparelli as canals. Drawings by W. R. Dawes (1865) also showed lineaments. The big question was whether or not these features were true lineaments or merely tricks of the human eye in joining together faint shadings and markings into linear features. Typical features mapped by Proctor are shown in this drawing.

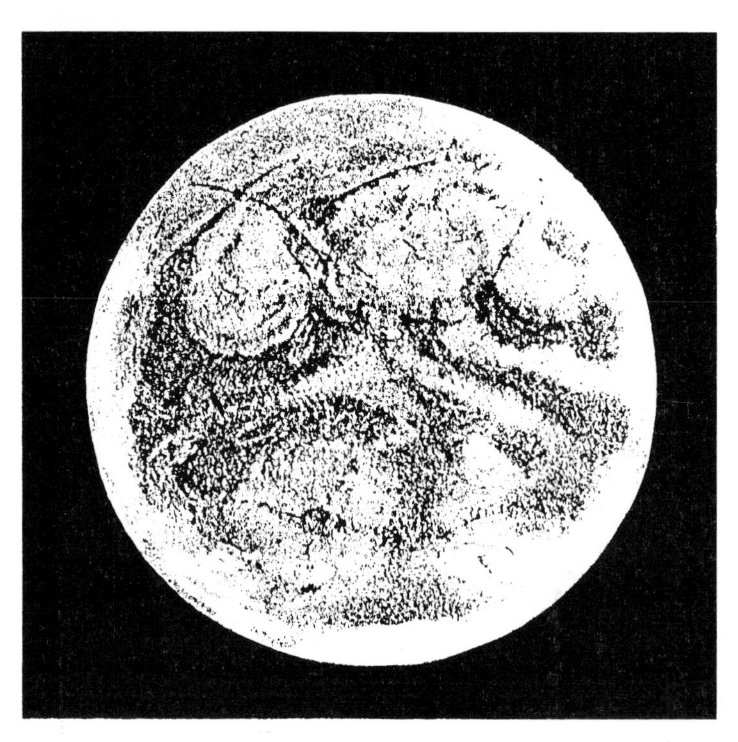

FIGURE 1.6: Mars seen through the large equatorial telescope of the Observatory of Meudon, near Paris, showed no trace of the fine lines when the planet was mapped by the astronomer E. Antoniadi. This was the most powerful telescope in Europe at that time (1909). Drawings such as the one shown here revealed splotches of fine detail which the "canal" enthusiasts interpreted as lineaments.

thing. For the balance of the nineteenth century and the early years of the twentieth century Mars was the most popular planet for planetary astronomers and the general public. Even in the early years of the twentieth century, semi-popular astronomical books contained statements such as: "This goes far to prove . . . the truth of the theory of intelligent life on Mars." There was an era of specialists who concentrated on Mars; including astronomers such as Eugenios M. Antoniadi, William H. Pickering, Percival Lowell, W. F. Denning, James E. Keeler, Andrew E. Douglass, and Camille Flammarion.

In 1909, Antoniadi, working with one of the best astronomical instruments of that time, a 33-inch (83.8 cm) refractor at the observatory of Meudon, near Paris, was unable to see any fine lines on the planet, although his drawings of the surface included greater detail than any previous drawings (figure 1.6). What Schiaparelli had drawn as sharp lines appeared on Antoniadi's drawings as small spots, or as boundaries of regions of different shades.

The study of the channels, or canals as they became popularly known, reached its climax at the beginning of the twentieth century as a result of the activities and publications of Percival Lowell.

Percival Lowell was born just after the midpoint of the nineteenth century. He was introduced as a teenager to the fascination of astronomy when his wealthy father bought him a telescope. But it was not until middle age, after a career in business and as a diplomat in the Far East, that he directed his attention full time to Mars. He had succumbed to the excitement of Schiaparelli's discovery and the intriguing possibilities of there being highly intelligent beings on another planet. In the 1890s Lowell used his wealth to build an observatory in an ideal climate for the study of Mars, at Flagstaff, Arizona. Lowell had an able collaborator in the Harvard astronomer William H. Pickering, who had seen the fine lines on Mars at an earlier opposition and had explained them as strips of vegetation irrigated by waterways.

The idea of there being life on Mars was commonly accepted, and was a favorite subject for speculative articles in the popular press. The work of Lowell added impetus to this belief. While he was much criticized by many other astronomers and scientists, his public talks always drew huge crowds, and any announcement from the Lowell Observatory received widespread press coverage worldwide. People in the late Victorian age believed in intelligent life on Mars as strongly as many people today believe in super beings piloting UFOs. Lowell constructed a theory on the probability of life on Mars even before he started his systematic observations of the planet at Flagstaff.

Lowell was, of course, familiar with the observations and speculations of other astronomers and scientists. His observations reinforced all he had previously read or learned about Mars. He explained the lines as artificial, most probably canals constructed by highly intelligent Martians. But what these beings might look like he did not publicly suggest. It remained for the popular press to invent the idea of little green men from Mars.

While many people speculated widely pro and con the canals and their implications, Lowell based his theory on systematic observations of his own. He contended that systematic observations of Mars would produce empirical evidence not only about the canals but also about the inexplicable changes that occur on the surface. He developed the procedure of observing Mars whenever the planet was visible in the sky and to draw the same surface features over and over again, to record the changes. He emphasized the point of repeatedly drawing an area to eliminate an observer's personal biasses. He believed that if an observer did this systematically under the same climatic conditions to produce a series of drawings of a given feature on Mars the drawings would be scientifically comparable.

When photography came into use for astronomy he applied this tool too. He emphasized the need for many pictures being taken in the shortest possible time in the hope that one would be a good picture obtained while the atmosphere was momentarily steady. In May 1905 Lowell staggered even his critics with an announcement (*Bulletin* No. 21 of the Observatory) that he had succeeded in photographing the canals of Mars. This news headlined the front pages of many newspapers at that time, but unfortunately the canals were so faintly defined that they could not be reproduced by the techniques then available for newspapers and magazines. His staff had to spend many hours trying to fill the demands for prints from the negatives.

Earl C. Slipher arrived at Lowell Observatory in 1906 to continue the program of photography that had begun in 1901, lasted until the 1960s, and produced 126,000 images of the Red Planet. In 1907 the opposition was a perihelic one, during which Lowell spent his longest uninterrupted period observing at Flagstaff—a total of eight months. He photographed Mars in Arizona while Slipher photographed the planet from an observatory established in Chile.

Slipher claimed many photographs showed the canals, and in his book *Mars*, published in 1962, he included in each copy as plate XLVII a direct print from negatives (not a reproduction) of Mars which showed the lines on the planet corresponding to drawings of the planet by other astronomers. The controversy had by no means ended.

Lowell's globes and maps of maps were characterized by geometrical patterns of lines which he claimed were evidence of the presence of intelligent beings on the Red Planet. In retrospect we know that he also claimed similar markings on Mercury and Venus which have been shown by spacecraft to have no basis in reality. Although the spacecraft that traveled to Mars did not show these linear markings either, except for some gross features of canyons and rift valleys, the question remains as to why many people, including astronomers, of two generations should have firmly believed in canals on Mars, and intelligent life there.

Humans tend to be more excited about bizarre speculations than hard science. It is interesting to note that although 1907 was also a year of financial panic, *The Wall Street Journal* when summarizing the news at the end of the year stated that the significant news of 1907 was not the financial panic but the discovery by Lowell that there was intelligent life on Mars. People believed that the discovery was real. They argued about it, and talked about it, but the consensus was that there were intelligent beings on Mars.

Water on Mars was central to Lowell's theories of intelligent life there. So he set up a program to observe the planet spectroscopically when Mars was moving most rapidly toward or away from Earth. In this way he hoped to use what is referred to as the Doppler effect to separate the water vapor lines in the spectrum of Mars from those in the spectrum of the Earth. He did not do so.

While observations and interest in Mars continued through the twentieth century, hard facts were slow to accumulate. The son of a well-read Victorian who might have believed strongly in Martians would not know much more about Mars than did his father. He would have a smattering of information about its atmosphere but general speculations would be equally as wild about conditions on Mars and whether or not it was the abode of living things. He would be exposed to many novels about Martians in which he also might probably believe.

It was the next generation that experienced radical changes in human thoughts about Mars, changes based on hard irrefutable facts. During the lifetime of the Victorian's grandson our view of Mars changed significantly; in the decades of the 1960s and 1970s Mars was explored by machines from Earth. In place of the old controversies a new age of wonderment dawned. From a star in the sky to a world seen through a telescope, Mars changed into a place that mankind could visit, a world of many different places. The grandson saw clearly on his television that Mars does have channels, although not canals, and Mars does have extensive frozen oceans of deep permafrost as suggested by the Swiss writer Adrian Baumann in 1909. The pictures returned by spacecraft revealed that Mars has the biggest volcanoes in the Solar System, the largest areas of sand dunes, and the biggest canyons, and clouds and planetwide dust storms, and weather fronts.

Nevertheless, Mars is still the mysterious planet, but our natural wonderment has been expanded by knowledge. However the negative aspects of the cost-effectiveness of finding more about Mars replaced the insatiable curiosity of the Victorians who had expanded mentally into an intriguing universe. By contrast with the same period of the nineteenth century, toward the end of the twentieth century every action seems to demand a measurable economic return or physical pleasure if it is to be regarded as worth doing. Curiosity, for its own sake, seeking increasing awareness of nature to broaden individual human consciousness, no longer seems popular to people or to governments.

The change from Mars being a malefic influence to a world that mankind, through planetary engineering, might be able ultimately to adapt to the support of human life in

FIGURE 1.7: In 1952 the author published the first in-depth engineering study of an unmanned mission to Mars to map the surface of the planet in detail. The study originated the term "space probe" and showed that it would be possible to send unmanned missions to the Red Planet long before humans could go there. This picture by the author shows the Martian Probe orbiting Mars, gathering data about the surface, relaying it to Earth by radio, and collecting energy from the Sun by a plastic bag aluminized collector which drove a closed-cycle gas turbine electrical generator.

Martian colonies began in the 1950s. It would progress in stages. The first stage was the following two decades of intellectual, scientific, and technical preparation and actual missions to the Red Planet which culminated in 1977 with the landing of two *Viking* spacecraft on Mars. The second stage in the final quarter of the century was begun by the Russians; an attempt to explore the Martian satellites to be followed later by missions to the planet's surface by rovers, balloons, penetrators, meteostations and return of soil samples. A third stage, which also appears as though it will be led by the Russians, is expected to be the human presence on Mars in bases followed by permanent settlements. However, NASA's Office of Exploration has developed an in-depth plan for human exploration of Mars awaiting acceptance by the nation and by Congress. This is discussed further in a later chapter.

Man's imaginative notions of Mars as exemplified in the fiction of modern Martian mythology described in the previous section, gradually gave way to scientific notions which in their turn had to swing almost full circle from the Mars of Schiaparelli and Lowell to the Mars as revealed by spacecraft.

In 1952 the first published engineering study [E. Burgess, The Martian Probe, *Aeronautics* (November 1952), vol. 27], of a space probe to carry a payload to survey Mars from orbit and map the planet's surface (figure 1.7) had shown the technical feasibility of such a project and the way in which the Red Planet could be mapped by an automated spacecraft. Most of this technical paper was reproduced in a chapter of the book *Frontier to Space,* first published in England and the United States in 1955, which was translated into Russian and published in Moscow in 1957 just before the launching of the first Earth satellite, *Sputnik.* In the Russian description on the jacket was the following statement:

> The latest important questions which stand out for examination are the connected creations of artificial satellites not only of the Earth but also of Mars. Thus, for example, an artificial satellite of Mars will permit inspection of that planet within a small distance of space to revolve around Mars and will permit gradually to receive full, complete details of the surface of the planet clearly. This cannot be achieved with telescopes on Earth.

However, the intellectual climate was still inimical to general ideas of actual flights to other planets, so not until after the first satellites were launched in 1957 was serious consideration given to this idea of unmanned space probes to Mars, and it would be ten years after the *Aeronautics* paper before a Russian spacecraft would be the first spacecraft to be aimed toward Mars. A breakthrough in thought in the U.S. came in 1959 when a staff report of the Select Committee on Astronautics and Space Exploration of the U.S. House of Representatives stated: "Ever since astronomers first reported 'canals' and polar snow caps upon Mars, man has speculated about it. Is it a dead planet? Is it strewn with the remnants of ancient civilizations? Does life exist upon this world? For the first time now, a positive answer to this age-old riddle may lie ahead." But the first attempt to reach Mars was by Russian spacecraft. Two launches were attempted in October 1960. Both failed.

At the next opportunity for a Mars flight, October 1962, another Russian spacecraft launch failed, but in November 1962 the Soviets dispatched *Mars 1* to try to reach the Red Planet. The spacecraft commenced its long journey and its radio signals indicated that everything was working fine. But just before reaching Mars the transmissions ended mysteriously. A second spacecraft launched by the Soviets in November 1962 failed to leave Earth parking orbit for transfer into a trajectory leading to Mars.

At the time of the Soviet launching, the National Aeronautics and Space Administration established the Mariner-Mars project based on a spacecraft which NASA had successfully sent to Venus that same year. The project wanted to launch three spacecraft to the Red Planet. Actually only two were approved for missions because of Congressional budget cutting. The first U.S. spacecraft launched to Mars was in November 1964. Named *Mariner 3,* it failed immediately after launch. That November Russia also sent their second Mars probe, *Zond 2,* but it died on the way to the Red Planet.

Mariner 4, a backup to the ill-fated *Mariner 3,* was launched successfully. The first interplanetary spacecraft to reach Mars hurtled from the Florida launch site at 9:22 A.M. EST on November 28, 1964, atop an Atlas-Centaur booster. After its long journey, the spacecraft flew within 7400 miles of Mars on July 14, 1965.

But in selecting an imaging system for the spacecraft, imaging science teams had made a serious error of judgment. They had made no provision to obtain images of the whole disk of Mars during the spacecraft's approach to the Red Planet. The spacecraft sent only 22 disappointing close-up pictures back to Earth (figures 1.8 and 1.9). They were disappointing from the standpoint that nothing seen on them corresponded to views of Mars as seen from Earth. There were no pictures to show a transition from the Earth views to the close-ups provided by the *Mariner* camera. However, the *Mariner* pictures killed the Mars of mythology, and also the Mars of Schiaparelli, and Lowell,

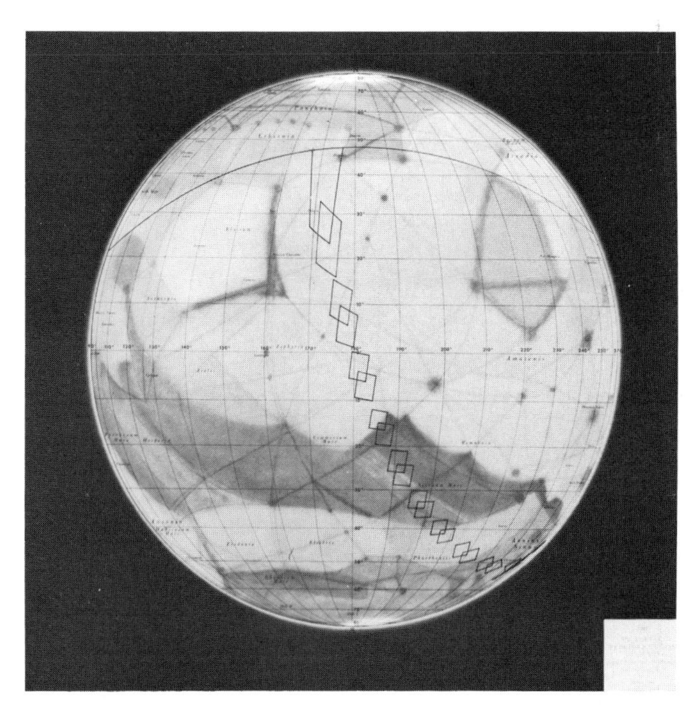

FIGURE 1.8: The first unmanned spacecraft did not reach Mars until 13 years after the author's Mars Probe paper. NASA's *Mariner 4* returned 22 pictures of the Martian surface in 1965. The areas covered by the 22 images are illustrated on this drawing. It is interesting to note that even at this time, official maps of Mars still included the lineaments of Schiaparelli and Lowell. (NASA/JPL)

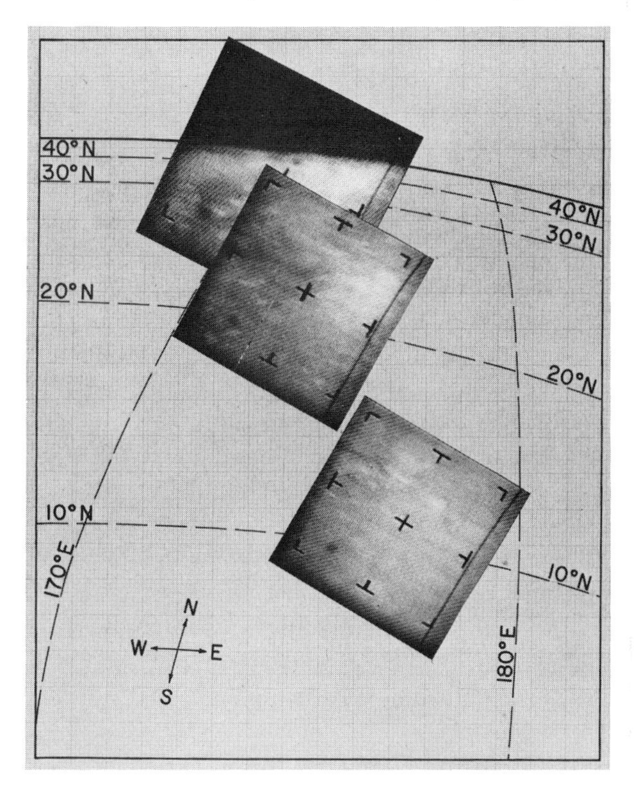

FIGURE 1.9: The first three rather disappointing pictures returned by the *Mariner 4* spacecraft are shown here in relation to the disc of Mars. (NASA/JPL)

and Burroughs, and of many astronomers too. The pictures revealed a moonlike cratered surface which was accepted as the real Mars by most scientists; a planet as dead as the Moon. Nevertheless, I pointed out in a paper presented at the New York Academy of Sciences and elsewhere that the pictures showed linear features that might be rift valleys (figure 1.10), might even be the features mistaken for "channels," and that the planet had been tectonically active—the surface had been molded by forces acting from within to deform the Martian crust, as well as by the impact of meteorites. This viewpoint was not, however, generally accepted, and the later Mars probes, *Mariners 6* and *7* in 1969, gave details that were recognizable as the albedo markings seen from Earth (figure 1.11), but in close-ups returned more views of a dead, cratered surface.

The 1969 *Mariners* stirred up a scientific hornet's nest. The old questions were replaced by new ones even more perplexing. Great stretches of Mars were heavily cratered (figure 1.12), yet no craters were visible on the one million square mile circular plain called Hellas. The surrounding cratered plains ended abruptly at the double scarps that formed the boundary of Hellas. The high plateau of Tharsis, a region of Mars that lacked features of interest when observed from Earth, displayed a complex pattern of bright curling streaks on the Mariner pictures, streaks that brightened each Martian afternoon.

Observations seemed to indicate that the bright areas of Mars were sparsely cratered and the dark areas heavily cratered—just the opposite to the Moon.

The nature of the bright polar caps sparked controversy. Were they water ice or solid carbon dioxide? Scientists could not be sure which because the evidence was still conflicting. Toward the edge of the polar caps, a white substance covered rims and floors of craters and where the sun had been shining on slopes the white deposit had disappeared to reveal dark material underneath. But deeper within the cap everything was covered with the white blanket. Sinuous rille-like markings were also discovered in parts of the caps.

However, the first Mars probes had only looked at one side of Mars. *Mariner 9*, which orbited the Red Planet in 1971, showed an intriguing new Mars, entirely different from the Mars which had occupied the mind of man previously. It was not the Mars of Lowell, neither was it the Mars of the earlier *Mariner* spacecraft. It was a Mars with vast chasms, enormous volcanoes, great dry river systems rivaling Earth's Amazon basin, of clouds and dust storms, and of remarkable layered polar caps surrounded by extensive dune fields. Mars was revealed as a world challenging Mankind to explore.

When *Mariner 9* arrived at Mars and entered an orbit around the Red Planet on November 13, 1971, scientists were chagrined to find the planet enshrouded in a globe-encircling dust storm. In the early approach pictures returned from *Mariner*, Mars looked like an old tennis ball with a white button of the south polar cap as the only feature visible through the dense global cloud. Several days later the pictures returned to Earth revealed four dark spots in the region of Tharsis. These proved to be very large craters, but it was several weeks before scientists admitted that they were calderas on the tops of great volcanoes, the biggest volcanoes in the Solar System (Figure 1.13).

Mars in the minds of men was changing again. It was now revealed as a planet on which there had been extensive periods of great volcanic activity. As the swirling dust clouds settled further exploration revealed the presence of the great canyons (figure 1.14), enormous dry river beds (figure 1.15), extensive lava flows, fault valleys, and layered ice caps. Scientists speculated that vast quantities of water must have flooded

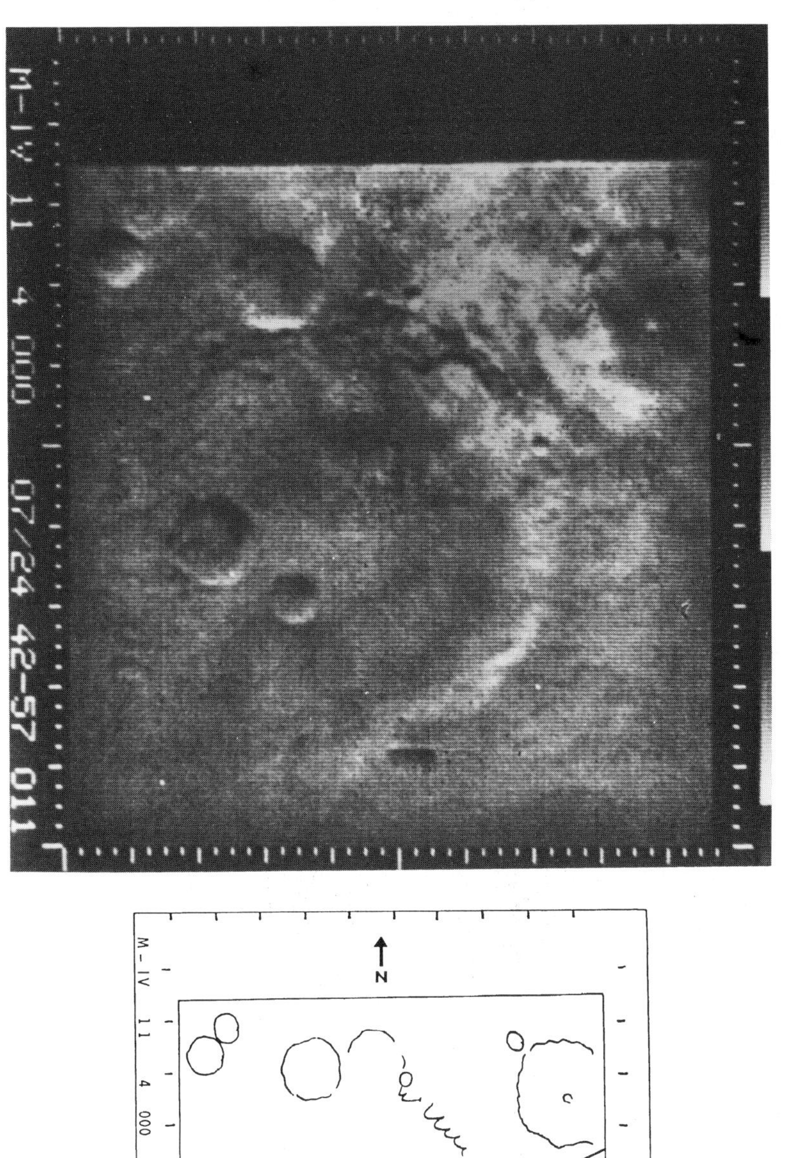

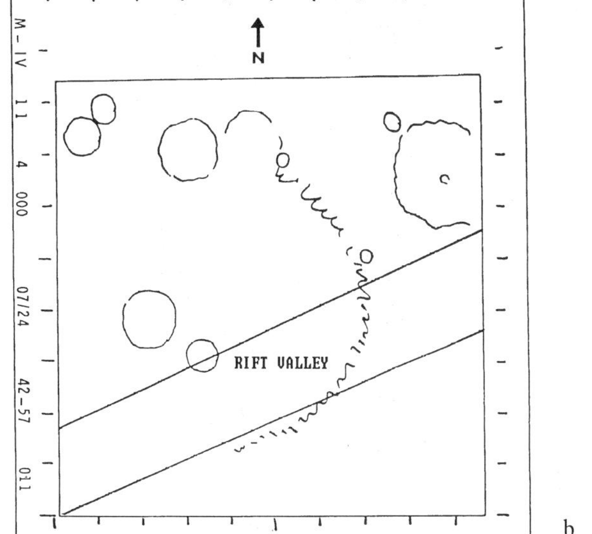

FIGURE 1.10: a) Like the others, NASA/JPL's picture number 11 showed a Moon-like heavily cratered surface. The idea of Mars as the abode of life appeared to have been vanquished. But the author discovered that this picture showed a lineament which indicated there had been tectonic activities on Mars, possibly lineaments that might explain the so-called "canals" as identified on the classical maps of Mars. This viewpoint was not, however, generally accepted. b) Line drawing to identify the rift valley which is now named Sirenum Fossae.

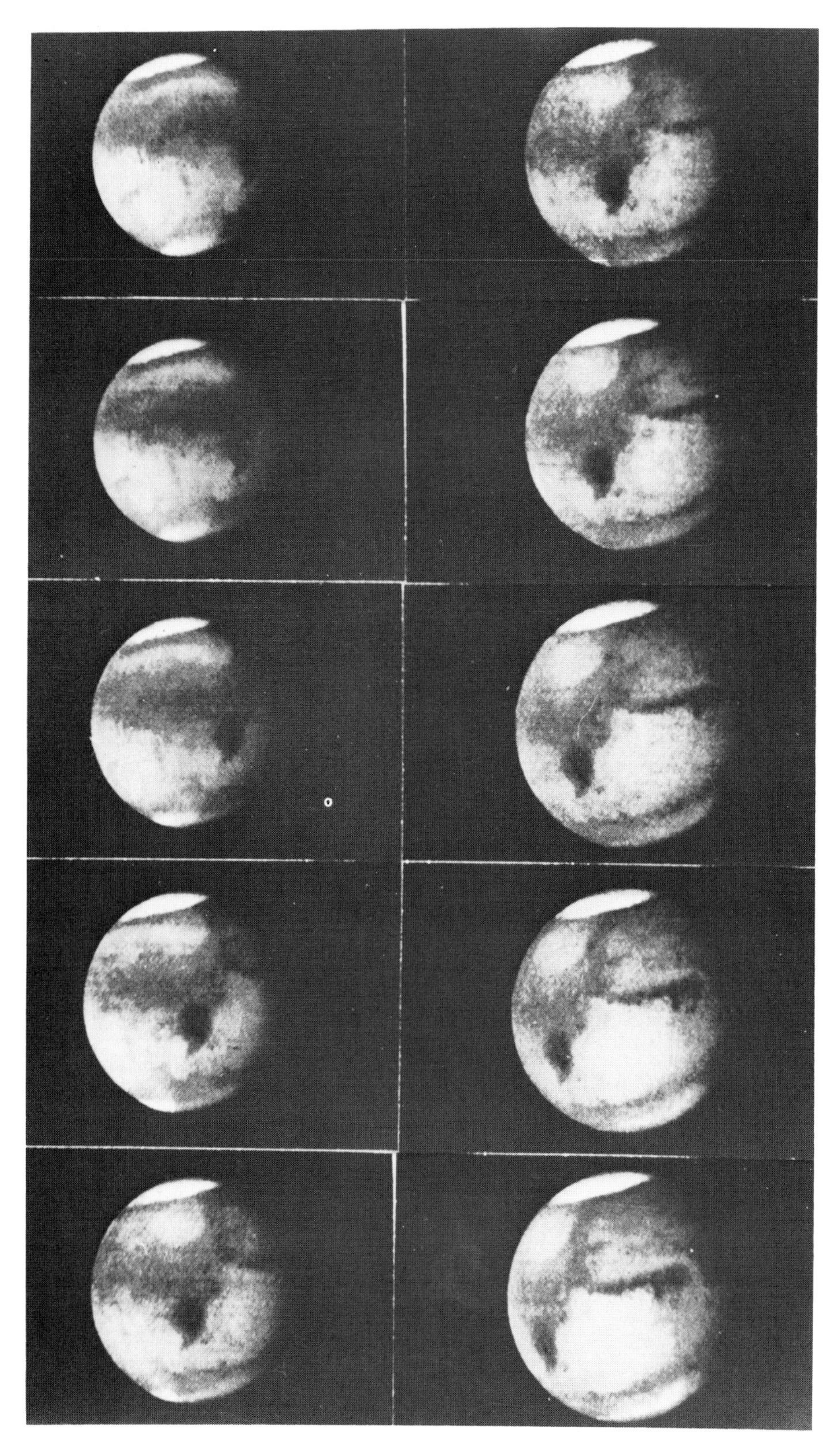

FIGURE 1.11: As Mariner 6 approached Mars in 1969 it obtained this series of pictures showing the rotation of the prominent Syrtis Major—a feature shown on Huygens' first telescopic drawing of the Red Planet in 1659. In this series Mars is oriented with the south pole at the top in the astronomical convention.

FIGURE 1.12: *Mariners* 6 and 7 again showed pictures of a cratered Mars as shown in this close-up from *Mariner 6.* Interest in the Red Planet waned. However, exobiologists still believed in Mars as the best candidate for life elsewhere in the Solar System. Orbiting and landing missions were planned as part of the continuing NASA program to explore the planets of the Solar System. (NASA/JPL)

over Mars in the past to account for the erosion and channels seen on the planet's surface. The layered ice caps indicated cyclic periods of great climatic changes on the Red Planet, and possibly times when the atmosphere was much more extensive than it is now. While today the pressure of Mars' atmosphere is about 1/100th that of the Earth's at sea level, scientists talked of periods when there might have been an atmosphere on the Red Planet with sufficient pressure to allow rain to fall on the Martian mountains, and rivers of water to flow into ancient ocean basins. Mars now began to appear as a world that had traveled partly along the evolutionary path toward becoming a world like the Earth and quite different from the Moon.

Meanwhile the science of exobiology, a search for life beyond Earth, had gathered many young scientists anxious to find universal laws about living things. The trouble with biological science is that it has only one form of life to observe; Earth life. So defining laws about living things can be a misleading exercise in futility without something other than Earth life on which to test the validity of such laws. While the search for life on Mars received a big setback following the early *Mariner* flybys of the planet, the new Mars that emerged from the *Mariner 9* expedition generated renewed enthusiasm. While it was most unlikely that a volcanic eruption would have been observed, the absence of craters on the slopes of the big volcanoes showed that they were relatively young. It looked as though Mars was an active rather than a dead planet. And an active planet might harbor life!

With each successive flight to the Red Planet there was a mounting feeling that we were getting closer to the real Mars. The *Mariner* program had brought back increasingly more refined information about Mars and while old questions were being answered new ones of even greater depth and complexity were being raised.

Clearly there were no seasonal changes that could be attributed to the growth and death of vegetation as had been suggested by Lowell and others. The changes observed in the light and dark markings appeared to result from dust storms and seasonal winds removing surface material to reveal darker or lighter material beneath. It was quite clear

a

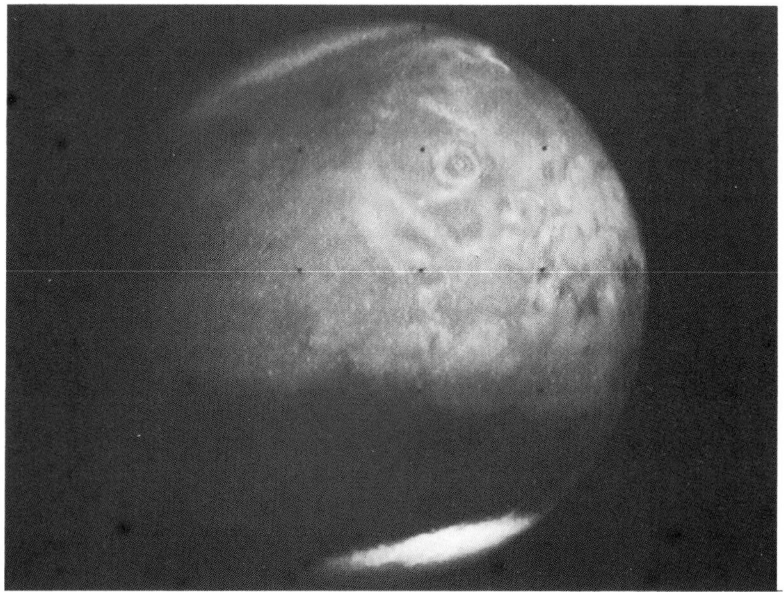

b

FIGURE 1.13: The Mars of *Mariner 9,* an orbiting spacecraft, was an entirely new Mars in the mind of man. The spacecraft discovered huge volcanoes on the Red Planet, including the mighty Olympus Mons which is the largest known volcano in the Solar System. a) Olympus Mons is identified as the pimple-like feature surrounded by a lighter colored circle on the *Mariner 7* picture. b) The volcano is seen in closeup from *Mariner 9.* While the volcano is now easily recognizable as such in the *Mariner 7* picture, it was a mystery at the time of the flyby because scientists did not then accept the possibility of a tectonically active Mars. In 1969, Mars in the mind of man was not a volcanic planet. (NASA/JPL)

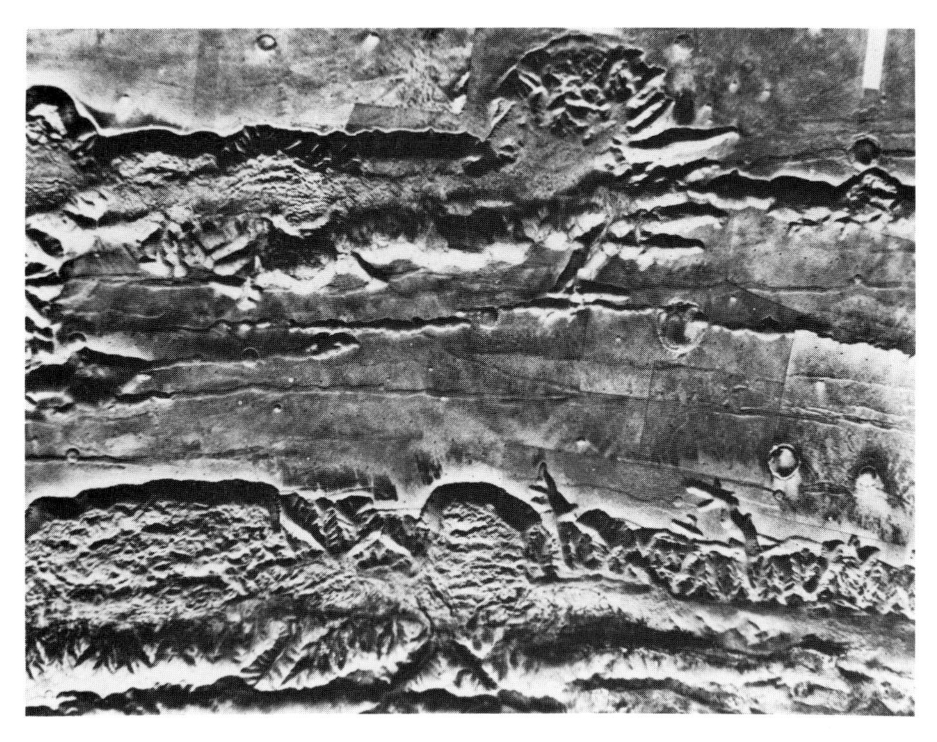

FIGURE 1.14: *Mariner 9* found great canyons that would stretch across the United States. This image shows two of the canyons with the tableland between them. Great landslides have spread debris across the canyon floors, there are fluted cliffs and sapping channels with tributaries leading into the main canyons. (NASA/JPL)

FIGURE 1.15: *Mariner 9* also discovered great dried river beds implying catastrophic ancient floods on Mars. The area shown in this Viking image is Northeast of Hellas. The valley cuts into layered deposits and the valley floor has a channel within it at some places. Note the fan of debris around the mountains near the center of the picture. (NASA/JPL)

that the planet had been volcanically very active. It was also clear that the thin carbon dioxide atmosphere of Mars could not support our type of life, nor was the pressure great enough for liquid water to be present on the Martian surface except possibly in the depths of canyons and the deepest depressed regions. We did not know if the volcanoes were still active, but we did know that there were regions of Mars where there might be sufficient water close to the surface to support living things.

The haunting question still remained about the possibility of life on the Red Planet today or in the past, and there was the exciting technical and philosophical challenge of searching for evidence to confirm or deny life on Mars.

The Space Science Board of the National Academy of Sciences recommended in 1965 that: "Specific life detection experiments should be incorporated in the early [Martian] landers." The exobiology study group of the Board had stated: "The biological exploration of Mars is a scientific undertaking of the greatest validity and significance." So the blessing of prestigious science groups had been given to the search.

Starting in the late 1960s, a plan to search for life on Mars was developed by NASA under the urging of exobiologists such as Joshua Lederberg, Nobel laureate geneticist, and Elliott Levinthal of Stanford Medical Center, Wolf Vishniac of Brookhaven National Laboratory, Gilbert Levin of Hazleton Laboratories, Norman Horowitz of California Institute of Technology, and Harold Klein and Vance Oyama of NASA-Ames Research Center. It is important to note that while in the Victorian age and for decades afterward it was individual scientists such as Schiaparelli and Lowell who pushed ahead the study of Mars, in the space age the push resulted from the concerted efforts of many scientists in different disciplinary fields, all anxious to know more about the Red Planet, including meteorologists and atmospheric physicists, geologists and planetologists, as well as biologists.

The search for life on Mars began with a program originally called *Voyager:* an ambitious undertaking to explore the planets including Mars, with a mission to the Red Planet planned for 1973. But Congress cut the *Voyager* program and NASA had to find other ways to reach and explore the planets. The decision of Congress to cancel *Voyager* reached the Martin Marietta Company, builders of the Martian *Voyager* spacecraft, on Labor Day weekend, 1967. It was an unexpected blow to many people who had devoted years of their professional careers to prove the feasibility of landing a spacecraft on Mars and to design this spacecraft. However, the engineers quickly recovered their enthusiasm and a new spacecraft was designed to suit the payload-carrying limitations of a smaller launch vehicle. A concept for a combination orbiter and lander was evolved that could be sent to Mars with a modified military booster—the Titan—instead of the big Saturn rocket required for *Voyager.*

By January 1968 the new project had been called *Viking* and had been approved by the President and by Congress. Two spacecraft were to be launched in 1973. Management teams were formed in April 1969 under the leadership of James S. Martin Jr., a veteran of the highly successful *Lunar Orbiter* program. The Mars project began operations at NASA's Langley Research Center, Virginia. *Viking* was the most complex unmanned space mission attempted so far (figure 1.16), requiring four spacecraft to be operated at the same time; two in orbit around Mars and two on the surface.

Once again Congress threw a monkey wrench into the highly efficient and carefully planned U.S. space program. The unique, advanced technology machine being developed for the exploration of Mars was set back by an indiscriminate slash to its budget in 1970. The cut was not great when compared to other federal programs, especially to the military waste in Vietnam, but it was proportionately very large for *Viking.* The landings on Mars had to be rescheduled to the less favorable 1975 opportunity. This caused significant changes to the program, such as cutting the amount of payload that

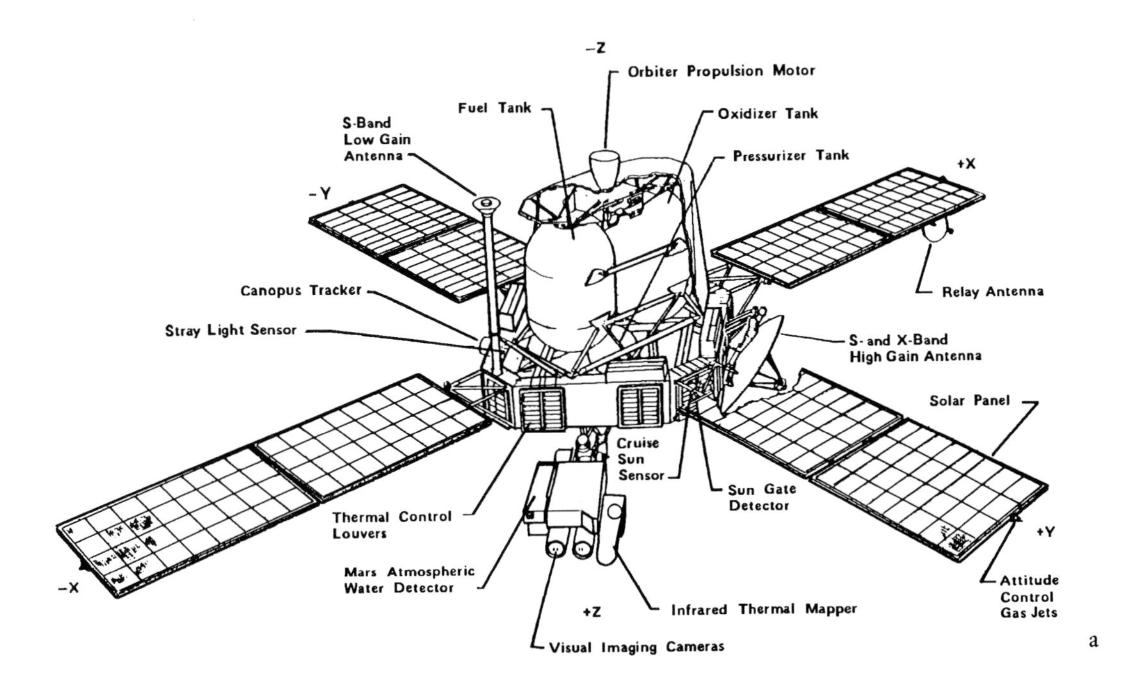

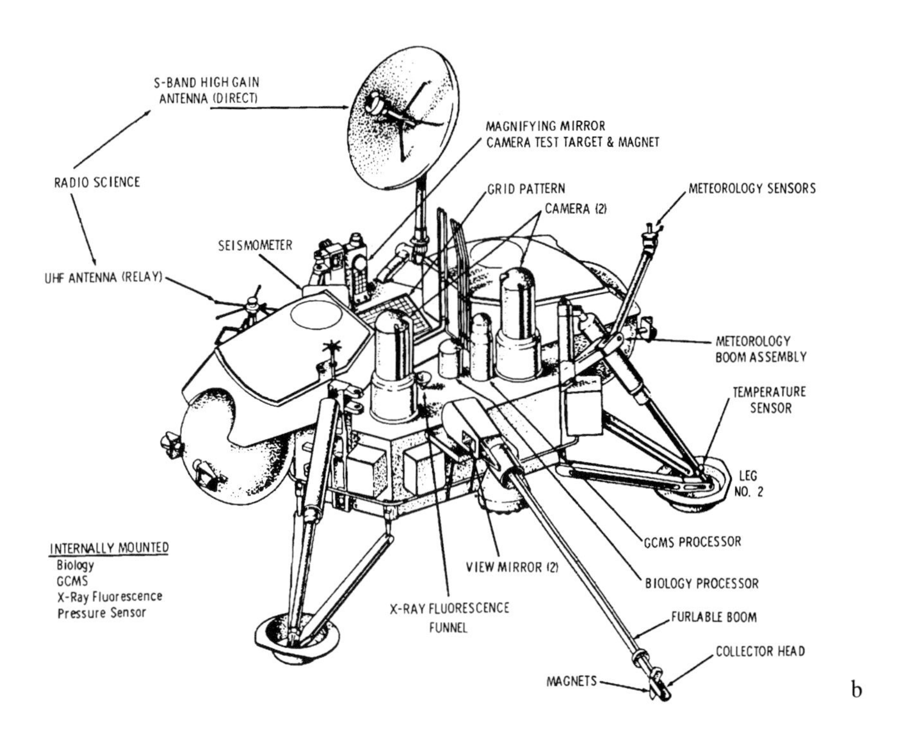

FIGURE 1.16: The *Viking* spacecraft, a combination of orbiter and lander, was the most sophisticated mission to explore another planet. The lander was carried below the orbiter on the journey to Mars, encapsulated in a bioshield to prevent it from contaminating Mars with Earth bacteria. The orbiter obtained electrical power from solar panels, the lander obtained its power from radioactive thermoelectric generators. a) Drawing of the orbiter to show its components and science instruments. b) Drawing of the lander to show its components and science instruments. The lander is depicted as it was on Mars with its sample gathering scoop extended to pick up Martian soil. (NASA-Langley)

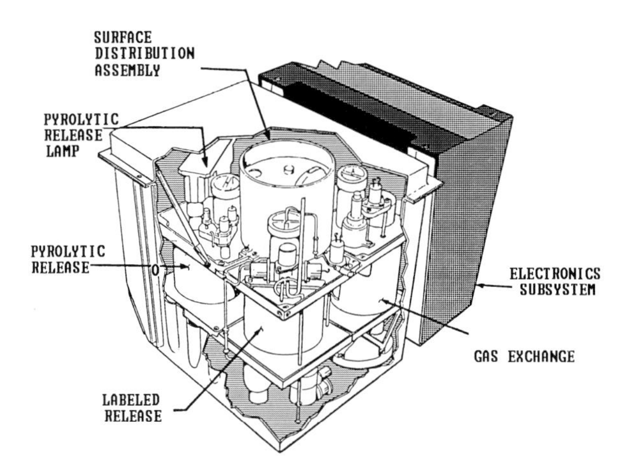

FIGURE 1.17: Complex biological laboratories were condensed to an unprecedented miniaturization to search for life on the Red Planet. This is a drawing of a test model of the biology package that was reduced in size to fit within one cubic foot.

could be carried and the amount of hardware available for the essential testing of the spacecraft and its systems before launch.

The biggest challenge of *Viking* was building the science instruments. Laboratories that on Earth traditionally took up the space of a whole room had to be packed into a cubic foot (figure 1.17). Another big challenge was to make the lander as the epitome of energy conservation. Anything not in use was designed to be automatically switched off immediately. All the scientific experiments, all the digging for samples on the Martian surface, and their analysis to search for evidence of Martian biology, and all the transmissions of information by radio from Mars had to be done with an amount of electricity equivalent to that used in a typical desk lamp—only 75 watts. A scoop had to be designed that would reach out and gather samples of the Martian soil and deposit them into the intake hatches of scientific experiments. An internal computer had to be heat sterilized so as not to contaminate Mars with terrestrial microbes; an unheard of torture for a digital computer. The lander had to be able to think for itself and guide itself down to the surface of Mars without human aid and then had to be capable of operating for months on that surface if for any reason the spacecraft could not be commanded from Earth.

Serious problems were encountered in developing the scientific instruments. Some types of soil clogged up the shutter that controlled its loading into one of the instruments. This was solved only 4 months before launch. A soil pulverizer ground unwanted metallic chips into the sample. This too was solved only just in time. Soil had to be delivered to three different biology experiments. Again the soil particles jammed metal surfaces sliding over each other. An engineer solved the problem by designing a stainless steel brush to sweep the soil samples into the test chambers.

The scientists tested the biological instruments in many different environments on Earth; snowcapped mountain peaks, tropical rain forests, dehydrated deserts. The instruments detected life in all the terrestrial environments, by enticing microbes to consume nutrients or to change the environment as a result of their metabolism. The exobiologists were confident that any microbial life on Mars could be detected by their instruments.

Big life forms would be searched for by taking photographs on the surface of the

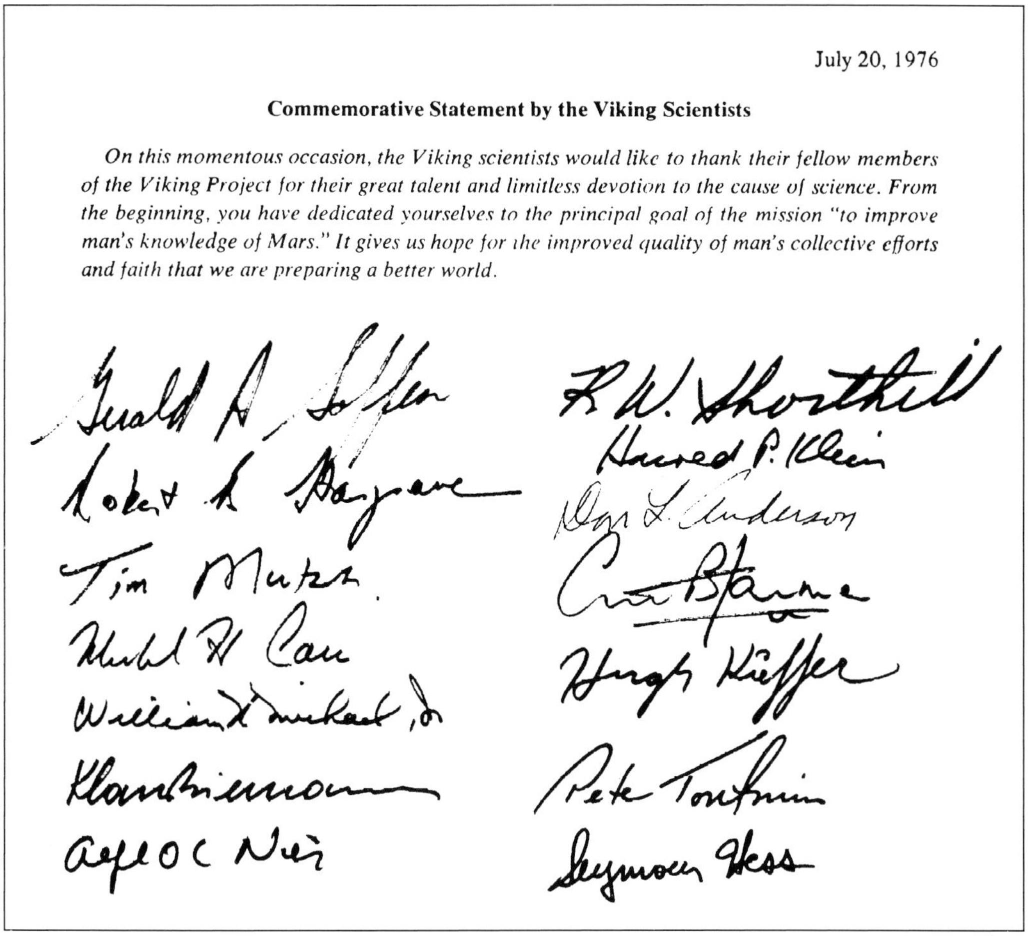

July 20, 1976

Commemorative Statement by the Viking Scientists

On this momentous occasion, the Viking scientists would like to thank their fellow members of the Viking Project for their great talent and limitless devotion to the cause of science. From the beginning, you have dedicated yourselves to the principal goal of the mission "to improve man's knowledge of Mars." It gives us hope for the improved quality of man's collective efforts and faith that we are preparing a better world.

FIGURE 1.18: The feelings about the importance of the exploration of Mars are expressed in this commemorative statement by the *Viking* scientists on the occasion of the first landing on Mars.

Red Planet. A special camera was designed to resist the expected sand-blasting of Martian dust storms. Each lander carried two of these cameras to produce stereo pictures of the Martian surface. If large living things were on Mars, and did not move about too quickly, they would be seen by these cameras. The search for life was both microscopic and macroscopic.

After a few delays at launch and some switching of hardware between spacecraft and of spacecraft between launch vehicles, both *Vikings* hurtled from the Florida launch site and were on their way to Mars by September 1975 scheduled to arrive at the Red Planet in June and August the following year.

An entirely new Mars was about to be revealed.

The new interest in Mars (figure 1.18) stimulated renewed speculation by a variety of people of the new generation. Articles began to appear more frequently in the popular press about Lowell and his theories. Speculation continued on what forms of life there might be on Mars. Even scientists waxed imaginative as they talked about creatures living underground or tapping underground water through deep root systems while protecting themselves from solar ultraviolet radiation by turtlelike carapaces. Scientific conferences heard papers on Martian life.

Science fiction magazine stories based on the findings of the *Mariners* held more closely to the real Mars. And because of the resurgence of mysticism and the occult, astral-projection believers were not idle either. In the tradition of Captain John Carter, the Burroughs hero, mystics attempted to project their egos across the millions of miles of space to obtain a view ahead of *Viking.* Several later recounted that they had received mental images of rocky plains and dune fields. And around the world people everywhere began to argue about the possibilities of life on Mars as ardently as during the era of Professor Lowell. The Red Planet had again seized the attention of the minds of men, and as *Viking* plunged toward its rendezvous on the Red Planet, hundreds of thousands of people waited expectantly, many perhaps secretly hoping that they would at long last see the little green men of another world, or at least the ruins of their ancient cities.

But this was not to be. Already the *Mariners* had shown that there was no evidence of artifacts of intelligent life visible from orbit, no ruined cities, no roadways, no vast areas of growing vegetation. The new Mars would bring a new loneliness to mankind. When the *Vikings* landed on Mars their message would be that our intelligent species and other terrestrial life forms appear to be alone in the Solar System. We have to rely upon ourselves to preserve our own future. There is no help from outside; no Martian records to tell us how to survive; no wise superbeings to solve our problems for us. The message from Mars would be that our survival is up to us. We are alone. But before the landing anticipation was still high that some form of life would be found on the Red Planet, even if only microbial.

It was the month of the U.S. bicentennial, a month during which *Viking* mission strategists had continued to diligently seek for and had finally decided on a landing place for *Viking 1* lander on the northern plain of Chryse which was possibly the dry bed of an ancient Martian ocean. Now a world waited expectantly for the first view from the surface of Mars.

Since the early morning hours on July 20, 1976, a steady stream of automobiles had flowed along Oak Creek Drive in the foothills of the San Gabriel Mountains, Southern California. Past Devil's Gate Dam and along the Arroyo Seco where experimenters from the Guggenheim Aeronautical Laboratory of the California Institute of Technology (GALCIT) had thirty-seven years earlier fired small rocket engines in first rudimentary attempts to reach for the planets as they developed rockets for the US Army. It was ironic that several of these pioneers of the space age were soon afterward forced to flee the United States and were lost to the U.S. space program during the incredible era of Senator Joseph McCarthy. Now, in 1976, at 4:00 A.M., the great complex of modern steel and concrete that was the Jet Propulsion Laboratory blazed with light at the head of the Arroyo as hundreds of space scientists, engineers and technicians prepared for an event that many had dedicated more than a decade of their lives to accomplish—the landing of a spacecraft called *Viking 1* on Mars.

Just three hours earlier a command to "go" for a landing had been authorized by James Martin, who had managed the National Aeronautics and Space Administration's *Viking* project since its inception in 1968. The combined efforts of thousands of people had paved the way for that command to start a complex sequence of events in the distant spacecraft which was then orbiting Mars and preparing itself for the landing. America's bicentennial expedition to the surface of Mars now reached its climax. And the miracle of electronics took the whole world along with the spacecraft. People at their home TV sets would be able to step onto the dusty Plain of Gold known as Chryse, an ancient Martian sea bed.

Mankind was about to step out and look across the landscape of an alien planet through an electronic extension of its human senses. *Viking* carried electronic eyes to see the red rocks and the alien skies of Mars; electronic ears to listen to Martian winds sweeping eerily across the Plain of Gold; and mechanical hands to gather the ancient soil of the Red Planet and feed it into automated laboratories that would be mystified by its unusual chemistry. The spacecraft even carried gifts of nutrients to entice any native microbial life form into revealing its presence.

The von Karman auditorium at the Jet Propulsion Laboratory, named after the California Institute of Technology professor who started the GALCIT rocket project in the late 1930s, was packed with representatives of the world's communicators—seasoned science reporters, magazine writers, television and radio commentators, and well-known names in science fiction, including Ray Bradbury. We were about to be on Mars. Not in the imagination of Edgar Rice Burroughs, or Ray Bradbury, or Arthur C. Clarke or the many other fiction writers about Mars, but really there on Mars through a series of technological miracles that had converted the fiction of yesterday into an even more exciting fact of today. Technological miracles were transforming a malefic god of war into a world of promise for the future of mankind.

Meanwhile the command to the *Viking 1* spacecraft had stimulated the machines of exploration into action, an action programmed months earlier before the machines left Earth atop a mighty Titan/Centaur rocket. The spacecraft that orbited Mars was initially two spacecraft, an Orbiter and a Lander. The Orbiter had carried the Lander some 360 million miles around the Sun since its launching from Florida 11 months earlier. A second Orbiter/Lander spacecraft, *Viking 2,* followed the first, and would reach Mars on August 7, 1976.

Viking 1 had reached the Red Planet on June 19. Since that time its cameras had been busily taking pictures of the surface looking for a suitable landing site. At first the search was unsuccessful, and a landing scheduled for July 4 had to be cancelled. But at last a suitably smooth area (figure 1.19) had been discovered in Chryse, a Martian plain.

At 1:50 A.M. on July 20, 45 minutes after receiving its go-ahead, the *Viking 1* lander began to operate as a separate vehicle. It separated from its orbiter and reoriented itself to fire its rocket engines, brake its orbital velocity, and enter a path leading down to Mars. At 80,000 feet above the surface the lander streaked into the Martian atmosphere. Its protective aeroshell glowed redly. Parts burned off as denser atmosphere at lower levels heated it further. Radio signals traveling at 186,000 miles per second from the spacecraft took 18 minutes to reach Earth, so the *Viking* actually landed on the Martian surface before its signals could reach Earth to inform its controllers that it had entered the atmosphere of Mars. Without human control the complex machine had to sense the atmosphere and feel its way down automatically to a soft landing on the distant planet.

As everyone waited for the signals to reach Earth, the chatter of conversation virtually ended in the auditorium. In offices throughout the space complex hundreds of people anxiously watched TV monitor screens on which a chart had appeared. It showed three curved lines traced across a graph of altitude plotted against time. Two curves set limits to the path of the *Viking* lander speeding toward the surface. If it descended too fast and did not automatically fire its rocket engines at the right time it would crash on the surface. If its computer fired its engines too soon all the lander's rocket propellant would be consumed before reaching the surface and that too would mean disaster.

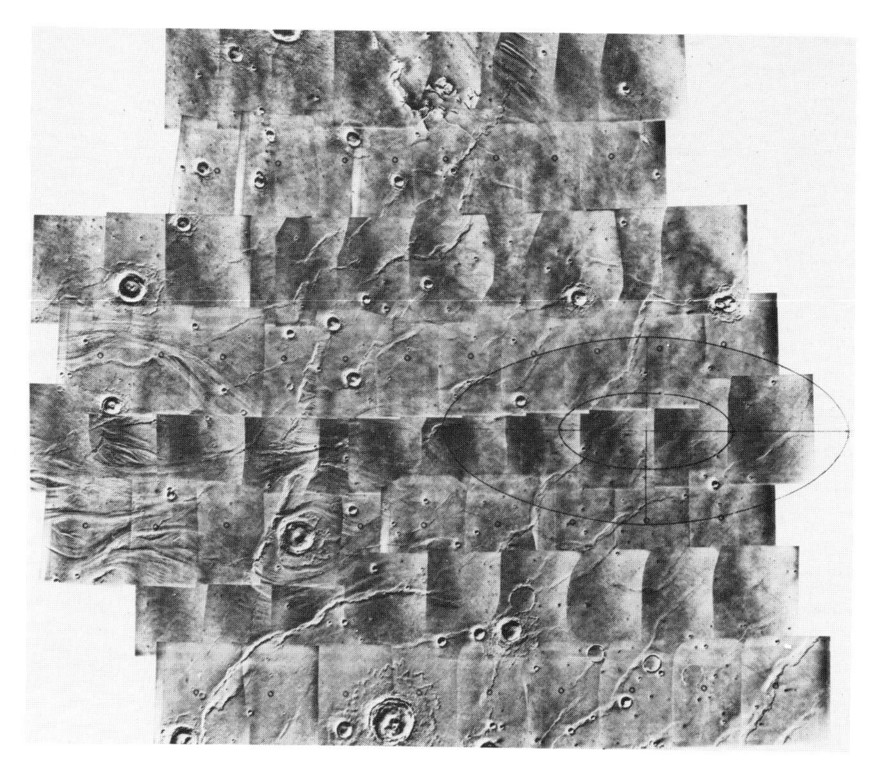

FIGURE 1.19: The plain of Chryse on Mars was selected for the first landing of a U.S. spacecraft on the Red Planet. The landing ellipse is shown superimposed on a mosaic of images obtained by the orbiting spacecraft, *Mariner 9.* (NASA/JPL)

Timing had to be just right so that the plots of the lander's position would always lie between the two limiting curves. A third curved line on the graph showed the path desired of *Viking 1* for the perfect landing.

Mission directors believed they had provided for every eventuality. But Mars had produced many surprises during past expeditions, and two attempted landings by Soviet spacecraft (landers from *Mars 2* and *Mars 3*) had both ended in disaster in 1971.

Small crosses appeared on the display, marking the actual path of the spacecraft. One by one they flashed on the monitor screens in the right position and showed that *Viking* was proceeding along the correct trajectory. At 5:09 A.M. telemetry confirmed that the parachute to slow the descent of the spacecraft had opened at the correct time (figure 1.20). Then came a signal telling of the parachute being jettisoned and of the terminal descent engines igniting as they should. Everyone was now quite silent, listening intently to the commentary and closely watching the TV monitors.

In the control center the eyes of Jim Martin watched words and numbers displayed on his monitor screen (figure 1.21). A data label showed ENABLET in the OFF position throughout the descent. Within one second of a successful touchdown this would change to ON. Another item on the display showed ENGON which signified that the electric heaters of the descent engines were ON. This would change to OFF immediately after touchdown. Yet another item showed voltage and current being used by the lander. These values would change abruptly at touchdown.

At 5:12 A.M. events moved quickly. ENGON flicked to OFF. ENABLET flicked to ON, and voltage and current levels changed.

FIGURE 1.20: This drawing by artist Don Davis depicts the opening of the parachute to slow the meteoric descent of the *Viking* Lander 1 in its plunge from orbit toward the plain of Chryse, an ancient ocean basin on the surface of Mars. Later still the parachute was jettisoned and the lander then used rocket engines to cushion it to a safe landing on the surface of the Red Planet. (NASA/JPL)

FIGURE 1.21: James Martin, project manager for *Viking,* watches his visual monitor screens in his office at the Space Flight Operations Facility of the Jet Propulsion Laboratory, Pasadena, California. He waits for the signals that will tell him that the first spacecraft has landed safely on the Red Planet. It took the signals traveling at the speed of light 18 minutes to reach Earth. The displays before Martin informed him among other things when the heaters on the descent engines turned off, and when the lander changed the rate at which it transmitted information over its telemetry link. These engineering data signals were the first vital signals confirming that the *Viking* had landed safely on the rocky plain of Chryse. (NASA/JPL)

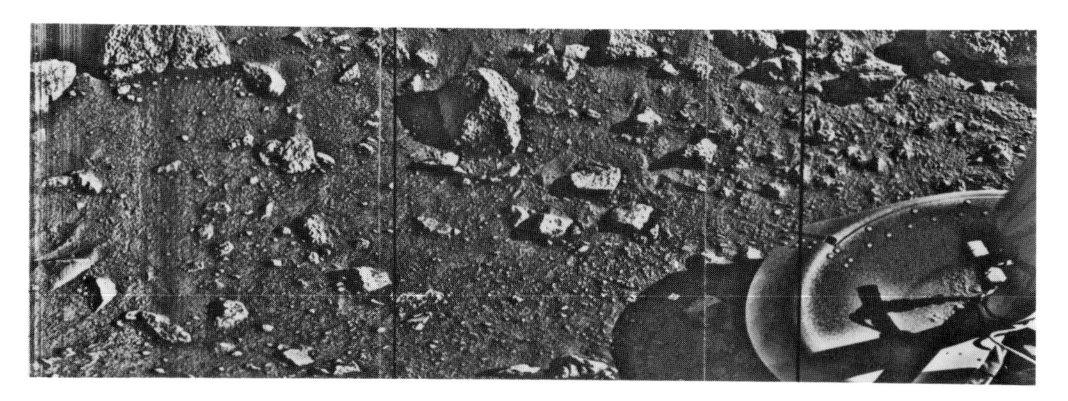

FIGURE 1.22: The first picture from the surface of Mars shows one of the footpads of the *Viking 1* Lander spacecraft resting on a rocky surface, quite unlike the surface of the Moon and more like a rocky pavement in one of Earth's deserts. (NASA/JPL)

FIGURE 1.23: The first panoramas of Mars were breathtaking; a strange world of rocks and dunes stretching across ridges to a far horizon with distant hills bathed in the light of the setting sun. The pink rocks emphasized the reality of a Red Planet. Again we were puzzled by the ubiquitous concept from antiquity that related Mars to a world of red fire and bloodshed. Especially when *Viking* discovered that even the Martian skies are red instead of the terrestrial blue. We do not have answers. And even though the idea of Mars as a god of war has been dispelled, there are new mysteries about the Red Planet that are equally as intriguing and thought-provoking as the ancient mysteries of the malefic planet that had puzzled mankind from the dawn of history to the time of the space age. (NASA/JPL)

"Touchdown! We have touchdown!"

Cheers erupted everywhere. *Viking* had not only landed on Mars but also it had landed safely and all its systems were operating.

The climax of the landing passed, but suspense again mounted as everyone waited for the first view of Mars to reach Earth. Would the *Viking* cameras reveal life; plants, shrubs? Would the surface be Earthlike? Moonlike? Would there be clouds? Would the sky be blue? What would we see on the surface of the Red Planet?

At 5:54 A.M. the blank viewing screens began to show action. Along the left side vertical stripes of varying intensity of light appeared, one after the other, slowly filling the screen. Gradually, a remarkably clear picture of the surface of Mars was revealed (figure 1.22) as though each of us were sitting on Mars on the lander and looking down at its footpad. Between rocks we could see a smooth soil, more like a desert on Earth than the meteor-churned regolith of the Moon.

A short while later a new picture began to arrive. This showed a panorama of Mars extending to the Martian horizon (figure 1.23). Different from the Moon, and more

TABLE 1.2. Missions to Mars

Name	*Origin*	*Launch Date*	*Arrival Date*	*Comments*
No name	USSR	Oct. 10, 1960	—	Failed at launch
No name	USSR	Oct. 14, 1960	—	Failed at launch
No name	USSR	Oct. 24, 1962	—	Failed in Earth orbit
Mars 1	USSR	Nov. 1, 1962	—	Lost communications
No name	USSR	Nov. 4, 1962	—	Failed in Earth orbit
Mariner 3	USA	Nov. 5, 1964	—	Shroud failed to detach
Mariner 4	USA	Nov. 28, 1964	July 14, 1965	First pictures of Mars
Zond 2	USSR	Nov. 30, 1964	—	Lost communications
Zond 3	USSR	July 18, 1965	—	Lost communications
Mariner 6	USA	Feb. 24, 1969	July 31, 1969	Many pictures of Mars
Mariner 7	USA	Mar. 27, 1969	Aug. 5, 1969	Many pictures of Mars
No name	USSR	Mar. 27, 1969	—	Failed at launch
Mariner 8	USA	May 8, 1971	—	Failed at launch
Kosmos 419	USSR	May 10, 1971	—	Failed at launch
Mars 2	USSR	May 19, 1971	Nov. 27, 1971	Orbited 3 months, Lander failed
Mars 3	USSR	May 28, 1971	Dec. 2, 1971	Orbited 3 months, Lander returned 20 secs of data from surface
Mariner 9	USA	May 30, 1971	Nov. 13, 1971	Orbited and returned many pictures for first global maps of Mars
Mars 4	USSR	July 21, 1973	Feb. 8, 1974	Failed to orbit but returned pictures
Mars 5	USSR	July 25, 1973	Feb. 12, 1974	Orbited 10 days and returned 60 pictures
Mars 6	USSR	Aug. 5, 1973	Mar. 12, 1974	Lander crashed
Mars 7	USSR	Aug. 9, 1973	Mar. 9, 1974	Lander missed Mars by 808 miles
Viking 1	USA	Aug. 20, 1975	Jun. 19, 1976	Landed July 20, 1976, operated for 5 years on Mars and over 4 years in orbit
Viking 2	USA	Sep. 9, 1975	Aug. 7, 1976	Landed Sept. 3, 1976, operated for nearly 4 years on Mars and 2 years in orbit
Phobos 1	USSR	July 7, 1988	—	Command error turned antenna away from Earth, all contact was lost, Sept. 1988
Phobos 2	USSR	July 12, 1988	Jan. 29, 1989	Entered orbit around Mars, obtained images of Mars and Phobos. Contact was lost, March 27, 1989; mission ended.

like the western deserts of Earth, a rock-strewn landscape stretched into the distance with some ridges intervening and some rippled dunes away to the left. Distant hills gleamed brightly in the glow of an approaching sunset on Mars. Humans had their first experience of what it would be like to stand on the surface of Mars and look around on another world.

The story of the results of the *Viking* expedition to Mars is part of the story of humans and of a quest to understand the role we play in the universe. As the sun set at the landing site in Chryse, the *Viking* lander prepared to start its scientific work that would continue for several years. When the results of this work were analyzed, a new picture of Mars began to unfold, the planet became a world we can understand rather than a light in the sky that we must fear.

2

A MYSTERIOUS AND MAJESTIC WORLD

In looking at Mars from outside, viewing it as the H. G. Wells' fictional Martians invading Earth might have first seen our planet, we would see Mars as a cloudy planet, not completely covered with clouds like Venus, but more akin to Earth with its seasonal clouds, although not to the same extent. That Mars has an atmosphere was known from the late 1700s when astronomers built telescopes with sufficient power to reveal frequent obscuration of surface features by mists, and the growth and decay of bright polar caps with the Martian seasons (figure 2.1). But the density and composition of this atmosphere remained a subject of controversy for almost 200 years. Before the exploration of Mars by spacecraft the only definitely known constituent of the Martian atmosphere was carbon dioxide, although by analogy with the Earth, nitrogen was assumed to be the major constituent. Neither water nor oxygen had been detected conclusively, and the pressure of the atmosphere at the Martian surface had been erroneously estimated as about 85 millibars (about one twelfth that of Earth's atmosphere).

There were many surprises when *Mariner* spacecraft flew by and orbited Mars and when *Viking* spacecraft orbited and landed on the Red Planet. They discovered that the atmosphere is predominantly carbon dioxide with only 2 or 3 percent of nitrogen and mere traces of oxygen and water (table 2.1). So Mars has an atmosphere with a bulk composition more like that of Venus than the atmosphere of Earth. There is, however, a great difference. While the surface pressure of the Venus atmosphere is 100 times that of Earth, the surface pressure of the Martian atmosphere is less than one-

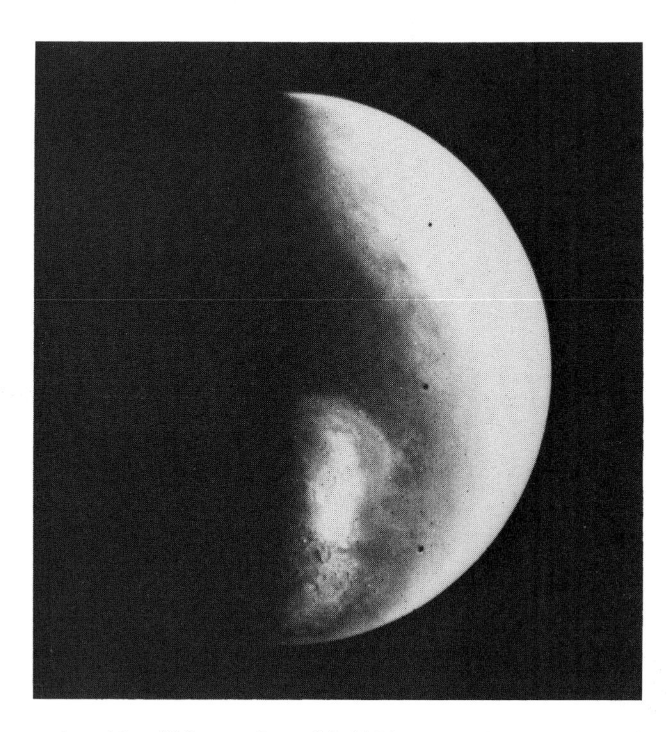

FIGURE 2.1: The morning side of Mars on June 16, 1976 as seen from *Viking 1* approaching the planet. Toward the bottom of the image is a very bright irregular feature within a somewhat less bright circular feature, the large impact basin of Hellas, which is about 1250 miles (2000 km) across. The irregular brightness is carbon dioxide frost. To the south of Hellas another brighter area is frost cover near the south pole which is in darkness to the lower left. (NASA/JPL)

hundredth that of the Earth; only about 7 millibars, or one-tenth of the estimate before spaceflight.

Other subtle differences are also present. Venus has sulfuric acid clouds and no ozone layer, which was quite unexpected. Mars by contrast does have an ozone layer like the Earth, which again was unexpected. Ratios of isotopes of rare gases such as xenon and argon differ from those in Earth's atmosphere. Isotopes of an element have different atomic weights but the same chemical properties. Their proportions in a planetary atmosphere tell scientists about how the planet probably evolved. Implications are that Mars had a massive early atmosphere which was lost during the evolution of the planet. Other isotope ratios indicate also that the atmosphere of Mars is more

TABLE 2.1. Composition of the Atmospheres of Mars, Earth, and Venus

Gas	*Mars* %	*Earth* %	*Venus* %
Carbon dioxide	95.32	0.04	97.0
Nitrogen	2.7	78.08	2.0
Argon	1.6	0.93	0.2
Oxygen	0.13	20.95	0.06
Carbon monoxide	0.07	0.15ppm	—
Water vapor (variable)	<0.03	<4.00	<0.03

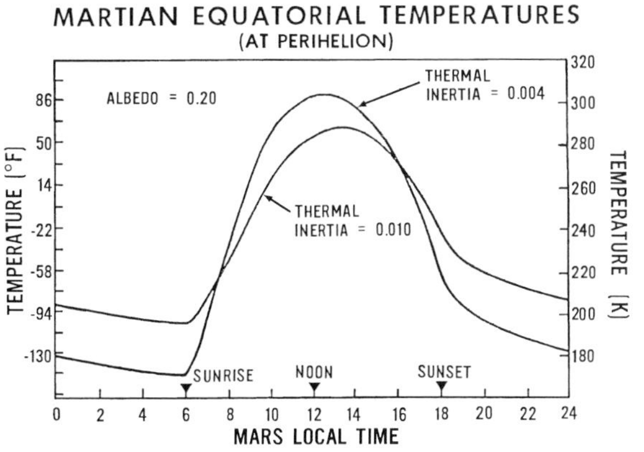

FIGURE 2.2: Variation of temperature on Mars in equatorial regions is shown for a typical Martian day. The two curves are for surface materials of different thermal qualities. Note how the maximum temperature, as on Earth, occurs during early afternoon. (NASA/JPL)

Earthlike than the chemical composition would suggest. For example, the ratios of heavy to light carbon and oxygen atoms are identical to those in Earth's atmospheric gases, but ratios for nitrogen differ, thereby indicating a different history and possibly much larger amounts of nitrogen on Mars in past ages. Some scientists believe that the escape into space of hydrogen, oxygen, and nitrogen from the early atmosphere of Mars undoubtedly played an important role in determining how the atmosphere of Mars evolved to its condition today. However, the precise nature of this role is not yet understood.

Meteorological packages carried by the two *Viking*s accumulated records of daily weather patterns at two sites on Mars; in Chryse near the equator and in Utopia at mid-northern latitude. These show considerable temperature variations between night and day on Mars and wind patterns that repeat on a daily basis. The pressure of the atmosphere varies in both daily and semidaily oscillations, and there is a seasonal decrease in general pressure as each polar cap builds up to its winter maximum by condensing carbon dioxide from the atmosphere.

Weather on Mars is monotonous compared with Earth's weather. During the Martian summer at the *Viking* sites, the highest daytime temperature in midafternoon was −31° C. Just before sunrise, the temperature had dipped to −86° C. At *Viking 1* site the soil temperature ranged from −90° C. to −10° C, and at the *Viking 2* site from −90° C. to −5° C.

The temperature of the Martian atmosphere varies with time of day (figure 2.2) as was expected; it is highest in the early afternoon local time and is lowest just before dawn. The tops of large volcanoes are very cold; on 80,000-ft (24,400-m) high Olympus Mons and the slightly smaller Arsia Mons the temperature is only −129 and −135°C respectively. Moreover, the maximum daytime temperature is at times twice the minimum night temperature on top of the mountains.

Three regions of Mars are colder than average; a large area extending east and west of the big volcanoes in Tharsis, a small area west of the volcanoes of Elysium, and the large area of Arabia. Warmer than average areas are located along the deep canyons of the Valles Marineris and in the Chryse Planun. Another warm area is along the southern edge of Isidis Planitia, an ancient impact basin.

The *Viking* orbiters provided global coverage of the amount of water vapor in the

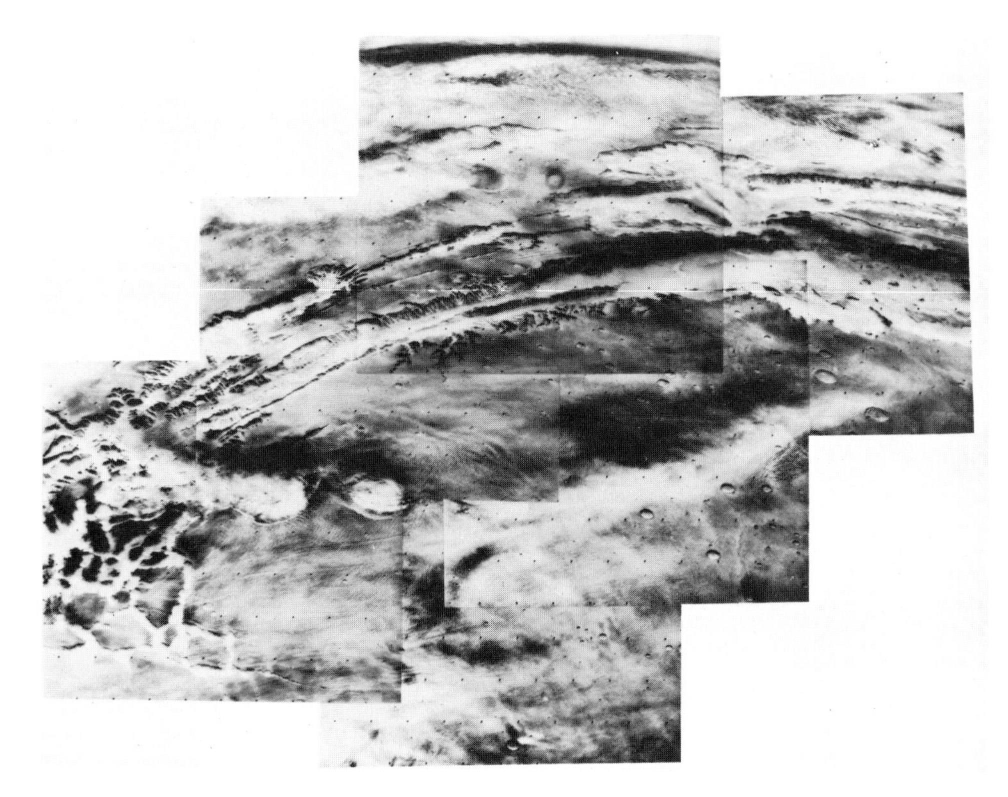

FIGURE 2.3: The great canyons of Mars are here filled with mist while diffuse white clouds spread over the surrounding plateaus. (NASA/JPL)

atmosphere of Mars. This amount ranges from zero at dawn to a maximum at midday. During most seasons and at most latitudes the atmosphere of Mars holds all the water it can for its temperature and pressure. There is, indeed, often 100 percent relative humidity. In some areas surface frost flashes into vapor after sunrise giving rise to increased amounts of vapor in the atmosphere which, in turn, produce morning mists.

Global concentrations of water vapor in the Martian atmosphere correspond to surface elevations; water vapor concentrates in the valleys and depressions of the Martian surface, and its concentration is lowest at the high mountains and volcanic areas. There is also a concentration in northern midsummer of water vapor in a low-lying belt surrounding the planet at high latitudes. Drainage regions for water vapor cover vast areas of Mars with at least two "water holes" where the amount of vapor is sometimes four times greater than in surrounding areas. One is located at the equator and another is outside but close to the great depression called Hellas. There are also two troughs of low elevation on Mars extending from the equator to the polar regions. One is Mare Acidalium, north of Chryse Planun, and the other extends north of Utopia. Both these areas appear dark on the classical maps of Mars. In these regions great changes in water vapor concentration are measured.

Viking also discovered that close to the poles the atmosphere becomes so saturated with water vapor that light snow falls on the polar caps. The total amount of water vapor near the polar regions requires that the temperature of the atmosphere close to the surface must be too high for the permanent polar caps to be frozen carbon dioxide.

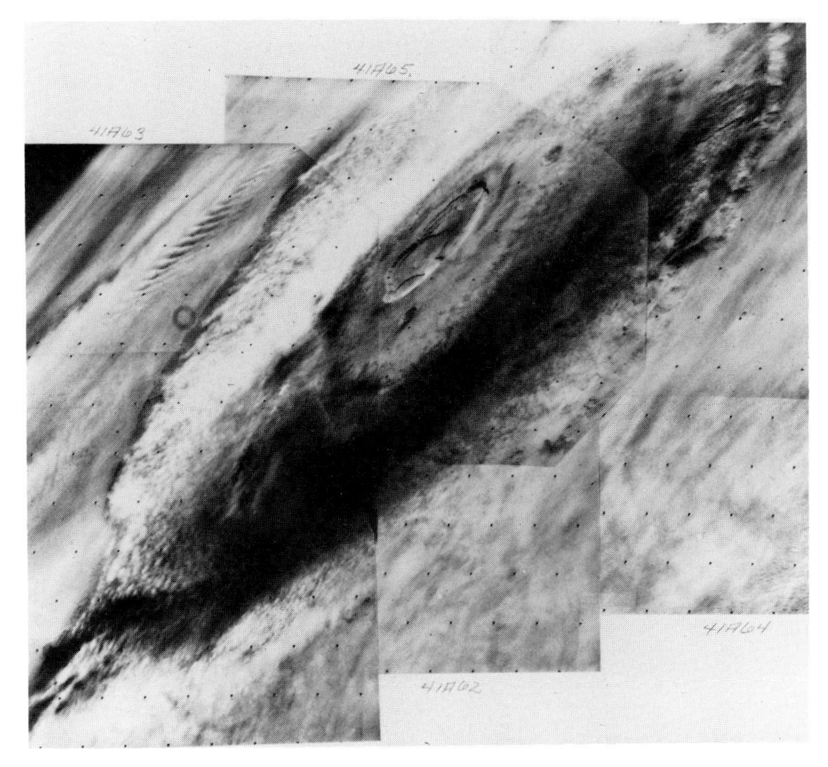

FIGURE 2.4: The 15-miles (24-km) high Olympus Mons is here seen at mid-morning on July 31, 1976, wreathed in clouds that extend up the flanks of the mountain to an altitude of over 60,000 feet (18,300 m). A well-defined cloud chain extends beyond the mountain in the upper left of the picture. (NASA/JPL)

They are mainly water ice. The more widespread caps that appear each winter are mixtures of carbon dioxide and water snows and frosts.

The Martian atmosphere has many different cloud forms; in fact, Mars is quite a cloudy planet. These clouds are present at scales ranging from tiny plumes streaming from peaks surrounding the calderas of the big volcanoes to extensive cloud systems. Some clouds move slowly, others quite quickly. Some clouds are obviously convective, similar to our cumulus clouds, others are dependent upon season and topography. There are fogs, mists, and haze layers. Meteorologists have identified at least nine clearly different types of Martian clouds.

Hazes are dense blankets of condensation that are more prevalent in the northern than in the southern hemisphere. They may be caused by condensation of water vapor around small particles of dust, and they often show wave patterns.

Diffuse white clouds appear in the haze blankets. They last for hours, sometimes for days, and have been observed from Earth for over seventy years. They do not seem to bear any relationship to topographical features on Mars. Thin wispy clouds close to the terminator which separates night from day on the planet look much like terrestrial cirrus clouds. The great canyons of Mars are often filled with mist (figure 2.3).

Large topographically related cloud systems are associated with mountainous areas, particularly the Tharsis plateau and the volcanoes. These clouds develop during the morning, first as diffuse masses and then into more complex structures. Figure 2.4 shows Olympus Mons wreathed in clouds that extend up the flanks to an altitude of

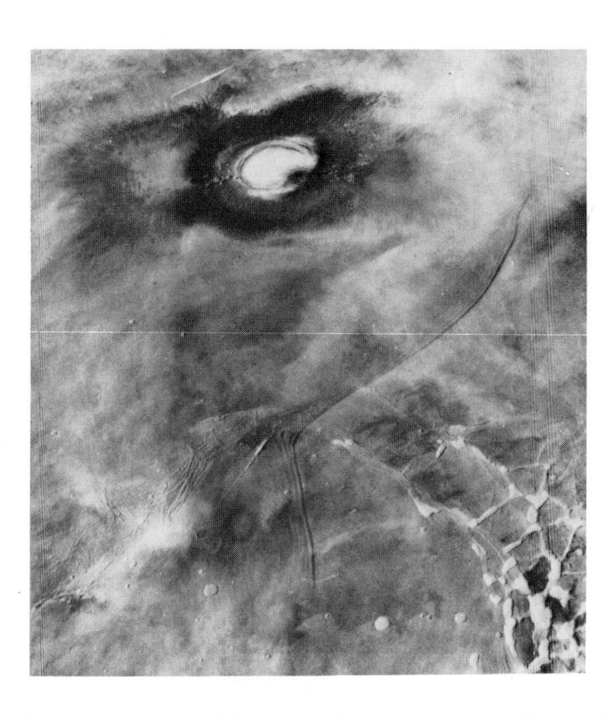

FIGURE 2.5: This oblique view of Arsia Mons and the western end of Noctis Labyrinthus shows the lower slopes of the volcano wreathed in clouds and extensive areas of mists covering the surrounding areas. Note the small cumulus-type clouds near the summit of the volcano. An unusual atmospheric feature appears as the curved line between Noctis Labyrinthus and the volcano. This is thought to be a weather front or some kind of atmospheric shock wave. Note the several waves of diminishing intensity in the lower part of the picture below the sharp bend in the line. (NASA/JPL)

over 60,000 ft (18,000 m). The cloud cover is most intense on the far western side of the mountain, and a well-defined cloud chain extends several hundred miles beyond the mountain to the upper left. There are also extensive layers of haze (figure 2.5). These clouds are large enough for astronomers to have observed them from Earth for many years. Some of the topographically related clouds are of water ice. The pictures from spacecraft show atmospheric wave patterns similar to those observed in the lee of terrestrial mountains.

Convective clouds are discrete cloudlets that do not change form but move as groups, sometimes at speeds of up to 140 miles per hour. They are optically very thin and float a few miles above the surface and do not seem to be affected by surface features.

Clouds in the region of Mare Acidalium, shown in figure 2.6, are typical winter clouds blanketing the surface from view. The bright circular feature marks the frost-rimmed walls of a crater, and the wind system has produced cloud wave forms on the lee of this crater.

Bright patches often seen in the bottoms of craters and channels are classed as local fogs (figure 2.7). They fill many valleys and deep depressions at times and are believed to be water vapor fogs. The featureless appearance of Hellas during the early flybys of *Mariner 6* and *7* resulted from its floor being completely obscured by a bright water vapor fog.

Detached thin cloud layers above the haze layers are seen in pictures of the limb of Mars (figure 2.8). The highest are 25 miles (40 km) above the surface. Essentially,

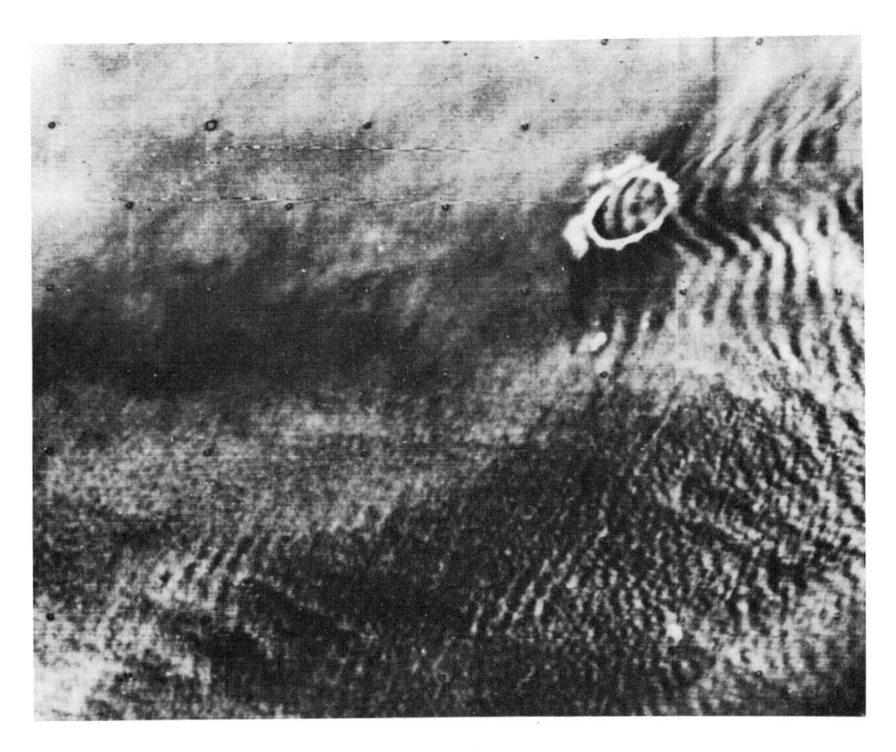

FIGURE 2.6: Typical winter clouds, like these in the region of Mare Acidalium, often blanket the Martian surface from view. The flow of air over the frost-rimmed crater has produced wave clouds which propagate for several hundred miles. Clouds of this type often occur on the lee side of terrestrial mountains. (NASA/JPL)

however, the layers would appear transparent to an observer on the surface of Mars looking up through them. Because the highest layers are warmer than the temperature at which carbon dioxide would condense, scientists say these cloud layers probably consist of water vapor.

Polar hood clouds of water ice, also observed from Earth, are very extensive in the northern hemisphere of Mars during the winter season. During winter the north polar cap is completely covered by such a hood. In early spring this hood disappears and the cap itself can then be seen. But even when the atmosphere is free of clouds there are still considerable quantities of water vapor in the atmosphere over the north polar cap (figure 2.9). In the northern polar regions, which have the severest winter and a large permanent ice cap, the amount of water available from the polar cap always seems sufficient to keep the atmosphere fully saturated there. In the southern polar regions, by contrast, the much smaller permanent cap cannot supply enough water vapor at the summer season to keep the atmosphere of the south polar region saturated. Water vapor then appears to move from the equatorial regions to the south polar region.

The atmosphere of Mars always contains suspended dust particles resulting from local and global dust storms that occur on the planet each Martian year. Many local dust storms have been identified in pictures returned from the *Viking* orbiters, and these storms frequently obscure surface details on Mars when the planet is observed from Earth. The global dust storms engulf the planet and prevent any surface details from being seen either from a spacecraft or through an Earth-based telescope.

a

b

FIGURE 2.7: Early morning fogs on Mars. Two pictures taken a half-hour apart by the *Viking 1* Orbiter show the development of early morning fogs in low spots, such as crater and channel bottoms (see arrows on bottom picture). Photograph at the top was taken shortly after Martian sunrise on July 24, 1976. The picture below was taken 30 minutes later. Slight warming of the sub-zero surface by the rising sun has driven off a small amount of water vapor which has recondensed in the colder air just above the surface to produce the fog patches. (NASA/JPL)

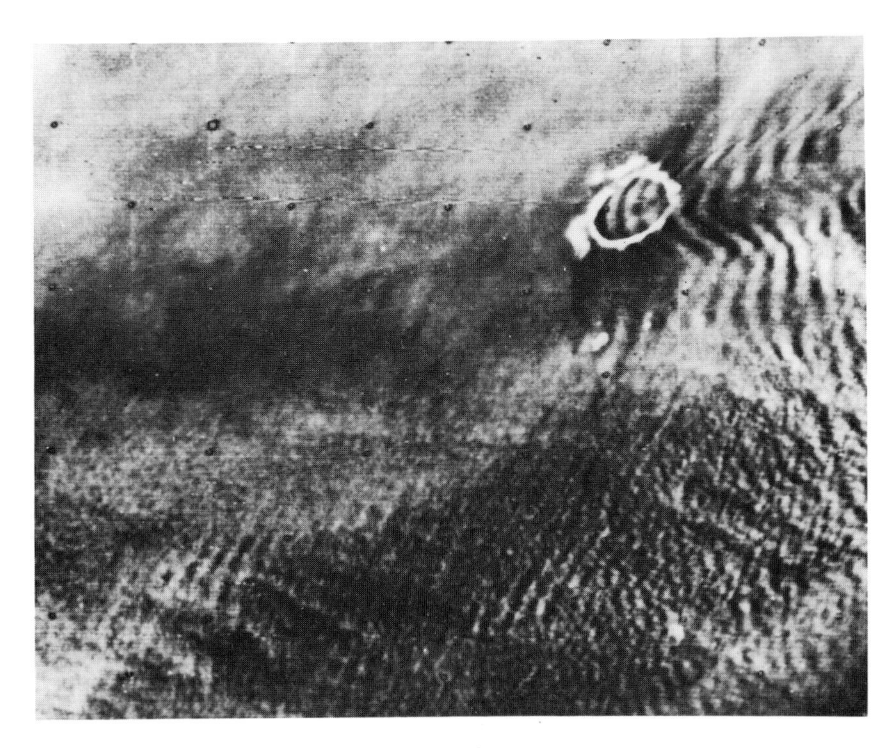

FIGURE 2.6: Typical winter clouds, like these in the region of Mare Acidalium, often blanket the Martian surface from view. The flow of air over the frost-rimmed crater has produced wave clouds which propagate for several hundred miles. Clouds of this type often occur on the lee side of terrestrial mountains. (NASA/JPL)

however, the layers would appear transparent to an observer on the surface of Mars looking up through them. Because the highest layers are warmer than the temperature at which carbon dioxide would condense, scientists say these cloud layers probably consist of water vapor.

Polar hood clouds of water ice, also observed from Earth, are very extensive in the northern hemisphere of Mars during the winter season. During winter the north polar cap is completely covered by such a hood. In early spring this hood disappears and the cap itself can then be seen. But even when the atmosphere is free of clouds there are still considerable quantities of water vapor in the atmosphere over the north polar cap (figure 2.9). In the northern polar regions, which have the severest winter and a large permanent ice cap, the amount of water available from the polar cap always seems sufficient to keep the atmosphere fully saturated there. In the southern polar regions, by contrast, the much smaller permanent cap cannot supply enough water vapor at the summer season to keep the atmosphere of the south polar region saturated. Water vapor then appears to move from the equatorial regions to the south polar region.

The atmosphere of Mars always contains suspended dust particles resulting from local and global dust storms that occur on the planet each Martian year. Many local dust storms have been identified in pictures returned from the *Viking* orbiters, and these storms frequently obscure surface details on Mars when the planet is observed from Earth. The global dust storms engulf the planet and prevent any surface details from being seen either from a spacecraft or through an Earth-based telescope.

a

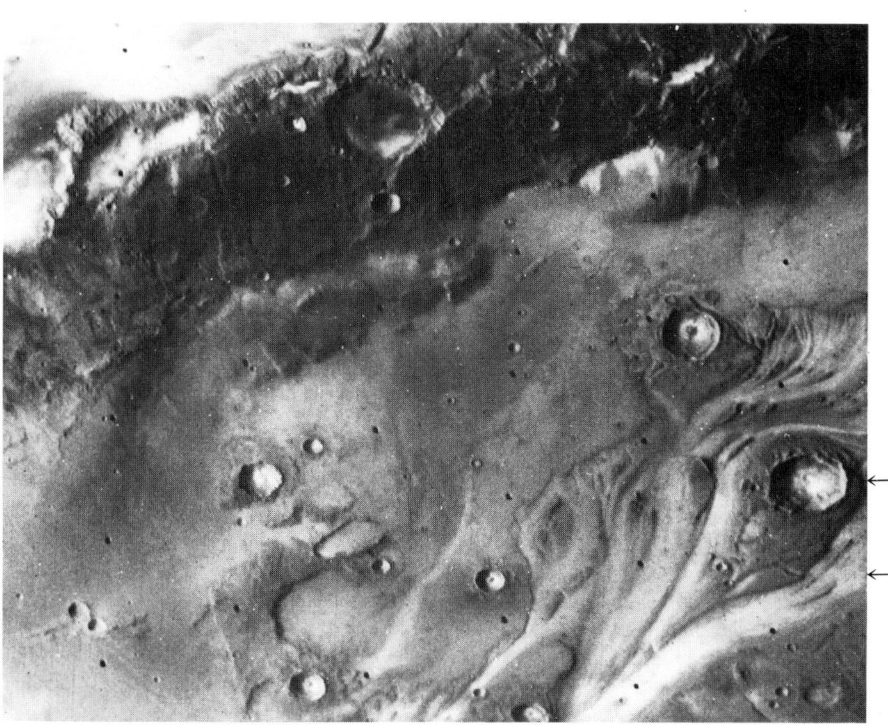

b

FIGURE 2.7: Early morning fogs on Mars. Two pictures taken a half-hour apart by the *Viking 1* Orbiter show the development of early morning fogs in low spots, such as crater and channel bottoms (see arrows on bottom picture). Photograph at the top was taken shortly after Martian sunrise on July 24, 1976. The picture below was taken 30 minutes later. Slight warming of the sub-zero surface by the rising sun has driven off a small amount of water vapor which has recondensed in the colder air just above the surface to produce the fog patches. (NASA/JPL)

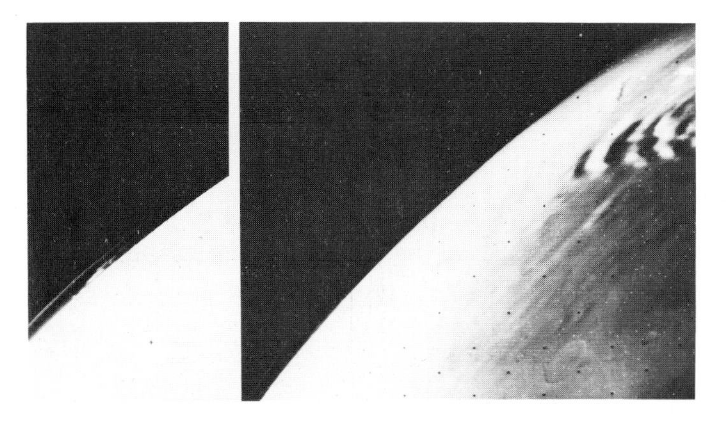

FIGURE 2.8: This picture shows high haze layers above the limb of Mars. These are clouds of ice crystals similar to high Earth clouds. The body of the planet is overexposed on this picture and appears totally white except in the right-hand picture which shows wave clouds while the limb hazes are almost invisible. (NASA/JPL)

FIGURE 2.9: The northern hemisphere of Mars—from the polar cap to a few degrees south of the equator—is seen in this mosaic of three photos taken by *Mariner 9* on August 7, 1972. The north polar ice cap is shrinking during the late Martian spring, but had been obscured by clouds of water and dry ice crystals until just a few months before this picture was taken. At this season the northern hemisphere appears free of clouds except for cloud systems streaming to the left of the big volcanoes near the bottom of the picture. The volcanic mountain Olympus Mons (lower left) is 310 miles (500 km) across at its base. (NASA/JPL)

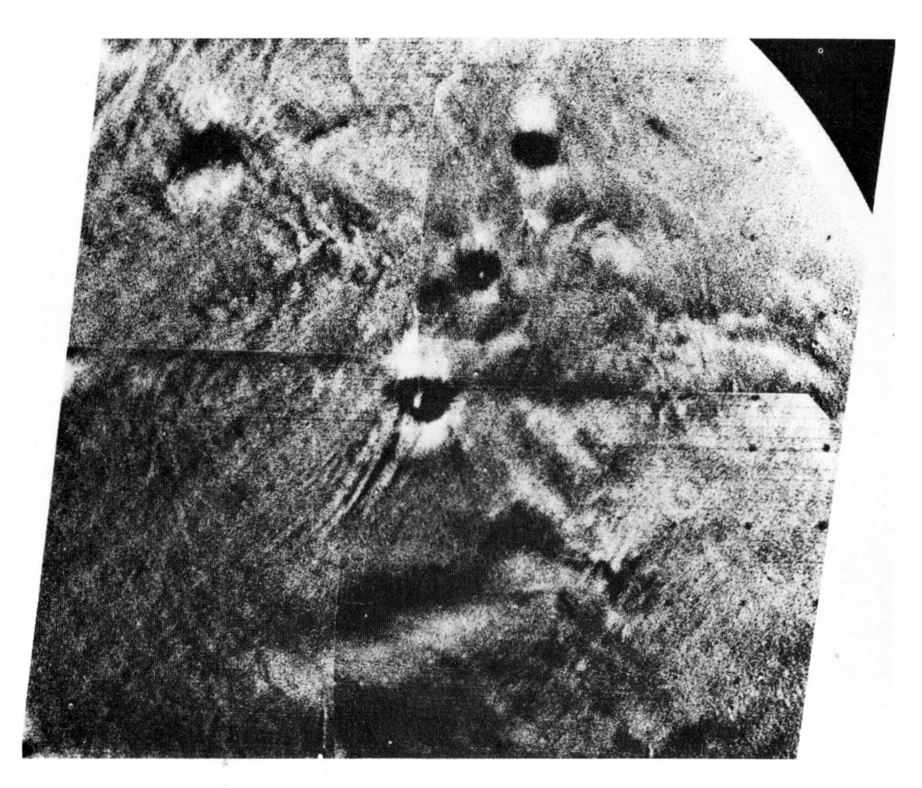

FIGURE 2.10: A planetwide dust storm raged on Mars at the time when *Mariner 9* went into orbit around the planet in 1972. The black spots are the tops of the four largest volcanoes in the Tharsis region. Olympus Mons is at top left of the picture. (NASA/JPL)

Dust storms are prevalent on Mars, as mentioned earlier. When *Mariner 9* arrived at Mars in November 1971 the whole planet was obscured by a dust storm with only the dark spots of several major volcanic mountains poking above the dust as shown in figure 2.10. The dust clouds stream downwind from the mountains. The *Viking* orbiter spacecraft also photographed many dust storms. A turbulent, bright dust cloud in February 1977 (identified by an arrow in figure 2.11) was more than 186 miles (300 km) across. It was located inside the great circular basin called Argyre produced ages ago by the impact of an asteroid, and is moving eastward under the influence of strong winds which are also giving rise to the wave clouds to the west of the basin (left of the photo). The lower half of the picture is brighter than the upper half because it is covered with carbon dioxide frost of the south polar cap. It is the end of the southern hemisphere winter on Mars and the south polar cap has begun to retreat. During the mid winter the whole of Argyre Basin (figure 2.12) is covered with frost.

The climate of Mars, similar to that of Earth, is produced by the inclination of the planet's axis to its orbit which once each year results in the northern and then the southern hemisphere more directly facing the Sun. Mars additionally experiences another important factor, a more elliptical orbit which moves the planet toward and away from the Sun. At present Mars is closest to the Sun and receives most solar radiation when the southern hemisphere is tilted sunward.

The response of Mars to seasonal changes is much more dramatic than is Earth's. Almost one fifth of its atmosphere condenses into polar caps each year, and the water vapor in the Martian atmosphere is at some seasons twice as much as at others.

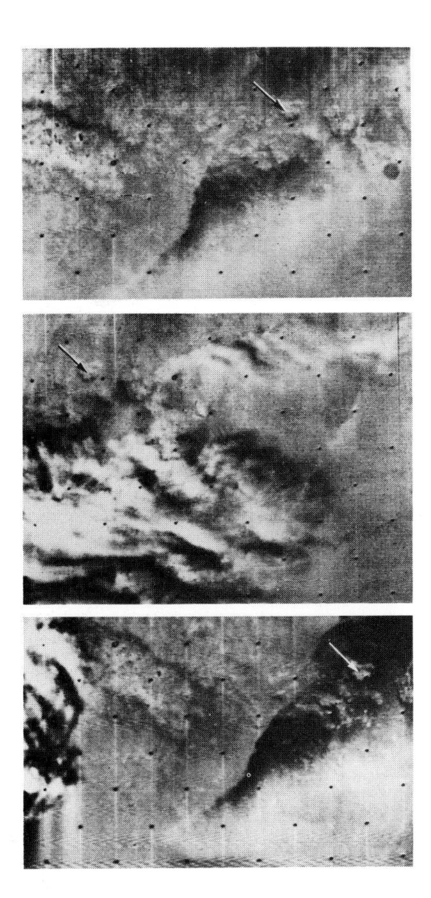

FIGURE 2.11: Typical dust storms on Mars. The top picture was taken February 11, 1972 in the northern region of Euxinus Lacus. The middle photograph was taken 24 hours later when the dust storm had spread to 300 miles (480 km). Two weeks later (bottom picture) shows the central area clearing of dust and a surface darkening because dust deposits had been removed from the surface. A new dust storm is at the left of the picture, about 190 miles (300 km) west of the first storm. Alternative covering and uncovering of dark surface by wind-blown, light-colored dust is believed to cause the changes in albedo markings that have been observed from Earth since the invention of the telescope. The arrows point to the same crater in each picture. (NASA/JPL)

Also, while terrestrial dust storms raised from deserts such as the Sahara can sometimes travel across the Atlantic Ocean to the Americas, generally Earth's dust storms are short lived. Dust is removed by rain and collected by the oceans. On Mars by contrast, dust storms can at times reach global proportions and shroud the whole planet in a ubiquitous and impenetrable veil.

A mystery about the climate of Mars is the tendency of water to concentrate in the northern hemisphere as though being pumped there by atmospheric forces. This could be the result of hotter spring seasons in the southern hemisphere subliming water more readily from the polar cap and generating strong winds that transport the water to the north more effectively than winds from the north in northern spring can move water vapor south.

Another phenomenon on Mars is that during the *Viking* mission carbon dioxide frost lasted all year round on the south polar cap, but disappears from the north polar cap in northern summer. This is mysterious because southern summers occur when Mars is closer to the Sun and would be expected to sublime all the carbon dioxide frost from

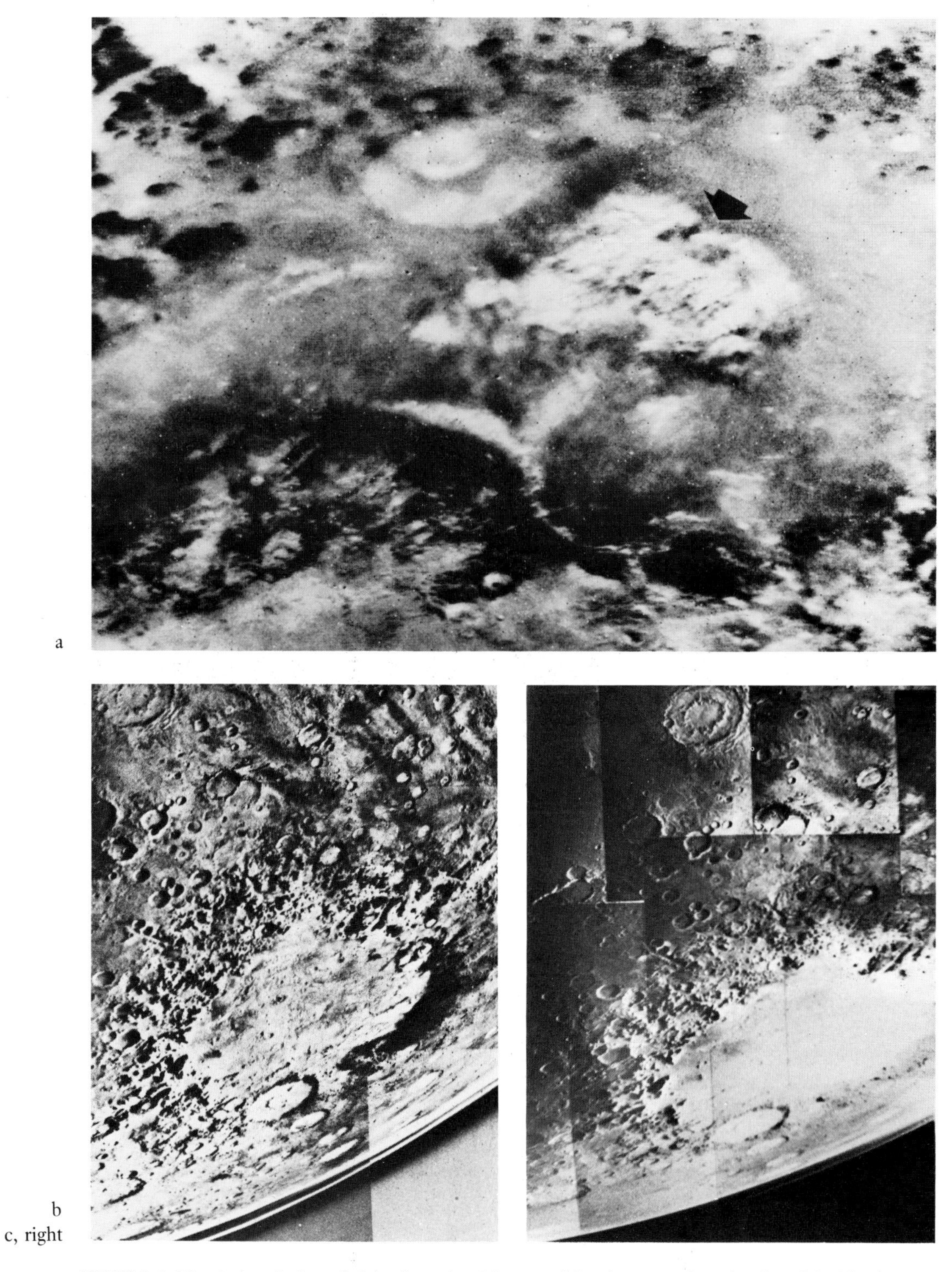

a

b
c, right

FIGURE 2.12: a) A turbulent, bright dust cloud indicated by the arrow has developed inside the great Argyre basin on October 2, 1977. It is moving eastward under the influence of strong winds. b) The great Argyre basin in the southern hemisphere of Mars is one of several enormous depressions created by the impact of large asteroids early in the planet's history. This picture shows it at summer season. c) In winter, the basin and the surrounding territory is covered with frost. (NASA/JPL)

TABLE 2.2. Composition of Martian Soil

SiO_2	Al_2O_3	Fe_2O_3	MgO	CaO	K_2O	TiO_2	SO_3	Cl
44	7	17	6	6	0	0.7	6	0.8
Rich	low	rich	low	low				

the south pole. A possible cause, which itself leads to other unanswered questions, is that the carbon dioxide frost on the south pole is brighter than that on the north pole. Thus it rejects more of the incident sunlight and is not heated enough for it to sublime. The darker north polar cap absorbs more heat from the Sun. Why the albedos differ is uncertain.

The upper atmosphere of Mars has an ionosphere where molecules and atoms of atmospheric gases are electrically charged by solar radiation. The major constituent of the Martian ionosphere is the single charged molecular oxygen ion at an altitude of about 80 miles (130 km). In this region carbon dioxide ions convert into the more stable oxygen ions. The upper atmosphere of Mars was found to be surprisingly cold compared with that of the Earth and to have a complex thermal structure.

From the standpoint of humans on Mars, the atmosphere cannot support life. Humans would need protection by spacesuits against low atmospheric pressure, low temperatures, reactive and abrasive dust, and against high energy cosmic rays and solar protons which the Martian atmosphere is too thin to stop from reaching the Martian surface. Such a spacesuit would require a life support system to provide oxygen and temperature control and to remove water vapor and carbon dioxide.

Soil is the interface between a planet's lithosphere and its atmosphere. Earth's soil is exposed to atmospheric weathering and to the effects of its biosphere. On Mars the soil may have been exposed to biotic effects in the early history of the planet; today, however, there appears to be a single soil unit on Mars, a unit covering most of the planet and thoroughly homogenized by dust storms and winds. For example, the soil was chemically almost the same at the two sites of *Viking* landers. The spectral fingerprints of the surface material are also similar over most of the planet.

Nevertheless, the soil of Mars is rich in information about the evolution of the planet and may be rich in information about the origin of life on any planet. For example, cementation in some places around the landing sites of the *Vikings* indicate soluble salts such as sulfates may be present in the soil.

The elemental composition of Martian soil analyzed so far and averaged over the two *Viking* lander sites in the northern plains of Chryse and Utopia is shown in table 2.2. We have, however, samples of only two sites on Mars, and we need to solve the puzzle as to why there is no organic matter at either site, not even as much as in the lunar regolith despite the influx on Mars of meteorites which contain organic material.

The fine material of the Martian soil consists of iron-rich clays, a small percentage (about 5 percent) of magnesium sulfate, and an equivalent small percentage of iron oxides. The *Vikings* did not directly detect oxidants in the Martian soil, but their presence is assumed to account for the results of the biology experiments.

A number of minerals have been identified in the Martian fines including palagonite, smectite, magnesium clay, and aluminum ferric oxides. The Martian clays are loaded with up to 17 percent of iron oxides and some magnetic minerals.

In terrestrial laboratory attempts to simulate spectra of the Martian soils, iron-rich clay samples came closest to matching the *Viking* results. In fact, the results of the pyrolytic release biology experiment were repeated with the iron-rich clay, montmorillonite, which in laboratory simulations gave results similar to those found in the Martian samples.

The large dust storms on Mars remove and deposit fine dust all over the planet's surface. However, local dust storms vary in their thermal and visual characteristics which may imply that the composition of the dust varies from place to place on Mars and is not homogeneous as was once thought.

About 100 dust storms occur on Mars each year, principally in the southern hemisphere, but most of these do not develop into planetwide storms of the types which develop when Mars is close to perihelion. Most of the local storms are relatively short lived and decay after a few days. Planetwide storms can last for several weeks and decay much more slowly. The short-lived storms appear to be associated with local heating where the Sun is shining overhead on the surface, that is, close to the sub-solar point, and also around the edge of the retreating south polar cap.

Viking orbiter observations showed that local dust storms appear to originate most frequently in certain areas of the planet, namely, the region surrounding Solis Lacus, the great Argyre and Hellas depressions, and Hellespontus. These are all in the southern hemisphere. In the northern hemisphere areas such as Chryse, Syrtis Major, and Cerberus produce dust storms also.

Planetwide storms that originated in the years 1922, 1956, 1969, 1971, 1973, and 1977 began in the region of the southern hemisphere between Hellas and Solis Lacus, with a whole group originating in Hellespontus, the western boundary of the 2.5-miles (4-km) deep Hellas basin (figure 2.13) which is about the size of the Caribbean.

A major question is how the known velocity of winds in the rare Martian atmosphere can raise dust which is thought to consist of particles of less than 30 micrometers diameter, similar to particles of clay. The winds are simply not powerful enough to do so. They lack sufficient energy to break cohesive forces between the particles. However, once particles have been raised into the atmosphere, winds of much less energy can keep them aloft. Several mechanisms have been proposed to account for the dust storms. If carbon dioxide and water becomes absorbed into the regolith there may be times when higher temperatures cause these gases to suddenly erupt from the soil with sufficient energy to push dust particles into the air and allow the winds to urge them into a dust storm. Another possibility is that the dust grains clump on the soil and the larger clumps are more easily moved into suspension in the air by the wind. A third possibility is that solar heating gives rise to circulating columns of air with a local increase in wind velocity. These rapidly rotating vortices give rise to dust devils of the type common in Earth's arid regions. Yet another possibility is the impact of larger, sand-grain sized, particles which could more easily be set in motion by the wind and would break the cohesive forces of the dust and allow it to be moved into the air by the same low velocity winds.

Dust appears to be removed on a seasonal basis from the dark areas and this appears to be a long-standing repetitive phenomenon since areas of dust removal such as Syrtis Major and Cerberus have been prominent features on Mars since the first telescopic observations of the planet. Figure 2.14 compares an area close to Cerberus where there are obvious changes in the dust pattern between the *Mariner 9* and the *Viking* missions,

FIGURE 2.13: Another enormous impact basin is Hellas whose floor is 2.5 miles (4 km) below the mean surface of Mars. This mosaic was the first showing the floor of the basin because it is most often obscured by mists and dust haze. There appear to be areas covered by clouds even in this picture. Many volcanic flows are on the southern rim. (NASA/JPL)

but the general form of the dark area remains much the same. There are also regions of dust deposition, such as Tharsis, Elysium, and Arabia, where deposits may have depths of 6 ft (2 m) or more. From thermal analysis of the surfaces of the big volcanoes, it seems that dust is swept downward to collect in layers as thick as 50 ft (15 m) near their bases. This dust obscures the relief, smoothing the surfaces around the lower slopes of the volcanoes. Near the summits, less dust is deposited and the volcano tops appear darker than their flanks. Dust is apparently being removed from their summits and accumulated on their lower slopes.

Dust is probably deposited evenly over the planet by the planetwide dust storms, but is subsequently removed by minor storms which transport it to the dust sinks in the bright areas.

Estimates of the thickness of the dust deposits based on thermal, radar, and visual observations suggest that the dust deposits are about 100,000 to 1,000,000 years old and are, in fact, deposited and removed cyclically. The change in orbital parameters of Mars might accordingly be responsible for moving the dust deposits from one hemisphere to another in cycles of this length.

The deposition of dust over geological time is expected to vary greatly as climatic conditions change. The eccentricity of the planet's orbit is thought to lead to episodes when the atmosphere of Mars varies greatly from what it is today. These changes are thought to cycle in periods ranging from 50,000 to 1 million years. When the atmospheric pressure is higher, Martian winds can move dust more effectively and dust

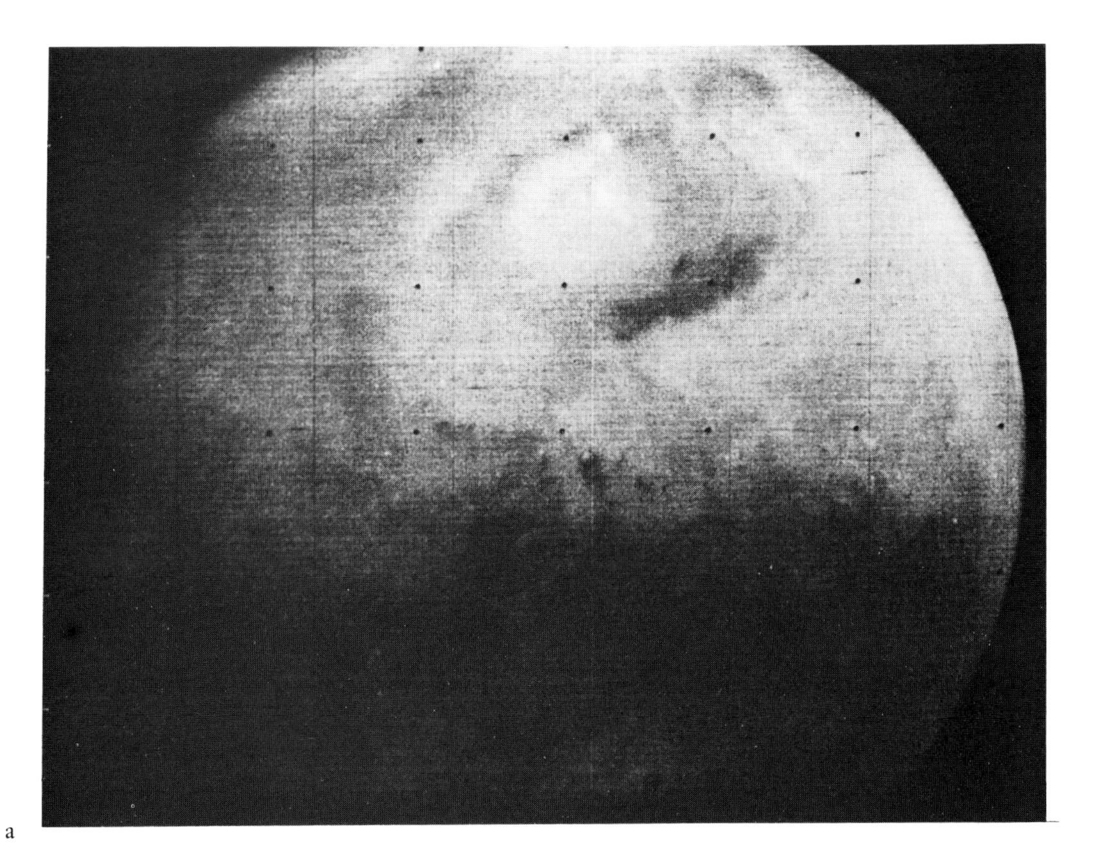

a

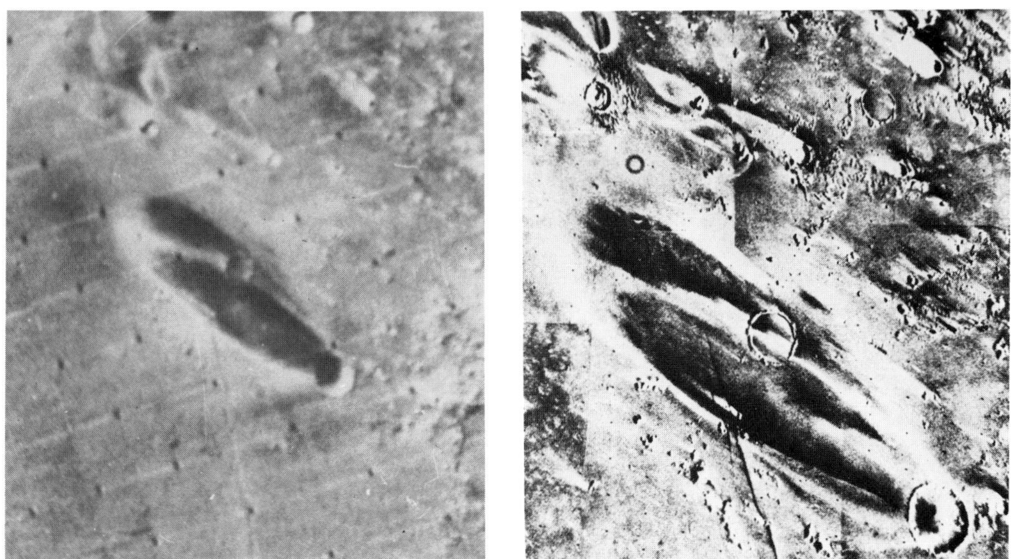

b
c, right

FIGURE 2.14: a) Cerberus is the elongated dark area below the white circular feature near the top of this view of Mars. The white feature is the Elysium volcanic region. Below and slightly to the west of Cerberus is a feature called Mesogaea where a 12-mile (20-km) diameter crater produces a persistent dark streak. b) This picture of the region was taken in 1972 by *Mariner 9*.c) A picture of the same area taken by *Viking* in 1976. There are subtle changes, but on the whole the dark pattern remains much the same. (NASA/JPL)

storms would be expected to increase. Should the density of the atmosphere be at times even less than it is today, there could be periods when Mars is virtually free of dust storms.

During northern hemisphere winter when the planet is closest to the Sun, dust storms are of two distinct types. Sometimes there are frequent local dust storms which only last individually for short periods but collectively produce a low altitude dusty layer of atmosphere surrounding the planet as a mid-latitude band. At other times dust storms develop in the southern tropical areas and from them the dust rises high into the atmosphere. Its subsequent heating by solar radiation creates strong circulating cells (Hadley cells) which rapidly cause a global storm pulling dust into the atmosphere everywhere.

At the present time, maximum wind velocities occur in the southern hemisphere when Mars is closest to the Sun and the southern hemisphere is tilted toward the Sun. It appears that the trend is now for dust to be removed from the southern hemisphere and deposited in the northern hemisphere of the planet, i.e., in Arabia, Elysium, and the region of the big volcanoes. When sufficient dust has been accumulated, a planet-wide dust storm can develop and redistribute the dust once again.

It now seems clear that the seasonal darkening of the dark (low albedo) areas of Mars is caused by the removal of dust from their surfaces. Dust is again deposited on them during major storms to be subsequently removed. Since major storms are seasonal, the albedo changes of the dark regions are also seasonal, thus completely disproving the old erroneous theories of the growth and death of vegetation each Martian year.

While dust and soil is moved around on Mars, actual weathering of rocks appears to be progressing at a very slow pace compared with past ages if based on the number of sharp craters in the areas of the *Viking* landing sites. It seems that most of the equatorial regions of Mars have hardly changed over many millions of years. These regions preserve ancient craters even down to small sizes. By contrast, large craters from the period of intense bombardment are all strongly eroded.

There seems firm evidence that an important role in the evolution of the surface of Mars is played by the repetitive episodes of dust erosion, transport, and deposition at various locations on the planet. Because the dust, once deposited is difficult to dislodge without some mechanism other than winds, its deposition over coarse sedimentary deposits and sands would have the effect of gradually trapping the dust and stopping the impacts of sand particles from raising dust clouds. The surface would, indeed, become choked with accumulated dust that would be difficult to dislodge. It appears that areas of Mars where this process is active today include the bright areas of Tharsis, Arabia, and Elysium, all fairly close to the equator in the northern hemisphere. During the big dust storms the dust appears to originate from the dark areas such as Syrtis Major.

That the dust originates from the dark areas is derived from observations showing that the atmosphere over these areas contains more dust than elsewhere even during periods when no actual dust storms are present. Also, the dark areas are darker during the period after a major dust storm, thereby indicating that dust has been removed from them during the storm. The fact that the dark regions remain visible virtually all the time indicates that the layer of dust on them must at all times be quite thin, probably only a few micrometers in thickness.

Not all the dark areas behave in exactly the same way. While many dark areas brighten and then fairly rapidly return to their lesser albedo (darker appearance) they had before a planetwide dust storm, Syrtis Major remains brighter than normal and is less distinct from its surroundings for a longer time, as does Acidalia on the far side of bright Arabia from Syrtis Major. These unusual areas probably also act as a source of dust for minor storms that occur between the global storms. At these times dust from the deep dust deposits cannot be taken up into the atmosphere, while dust from the thin layers on the dark areas can be dislodged into the atmosphere.

There are still many unanswered questions about the Martian dust, particularly regarding its source and how it is raised into the atmosphere, how it is removed, and how it originates. Earlier in this chapter ways of raising the dust with moderate winds were suggested. All have been studied in fair detail by researchers using theoretical models, wind tunnels, and observations at the *Viking* lander sites. It is believed that the dust originates from weathering of the Martian rocks which has been caused by volcanism, impacts of bodies from space, and chemical and mechanical interactions. On Earth dust is removed from the atmosphere by rain and by the oceans. On Mars it is probably removed by being deposited at the poles, giving rise to the laminated terrain there, and by trapping in rough terrain where it cannot be subsequently removed by dust devils or impact of sand grains.

Another difficult question to answer is why global dust storms do not occur each Martian year, even when the planet is close to perihelion. It could be that when there is plenty of dust in the northern band during northern hemisphere winter, sufficient stresses are not raised in the summer southern hemisphere for a global storm to be initiated. The converse would also be true; without strong dust bands in the northern hemisphere, the southern hemisphere could be triggered into a planetwide storm. The reason why the northern bands should vary is still unexplained. One suggestion is that following a global storm there is more dust available in the northern hemisphere for a strong dust band to develop and thereby prevent a global storm the following year, possibly for several Martian years.

It is important to note in connection with any plans for human exploration of Mars that the atmosphere of the planet is at all times contaminated by dust. Such dust may present a serious problem to life support systems, to mechanisms, and to spacesuits, and may contaminate the environment of the space vehicles and human habitats on the surface. In this connection it will be important to use hatch-entered spacesuits of the type now being designed for the American space station, *Freedom,* rather than spacesuits that have to be carried into airlocks. The latter could contaminate the atmosphere of the Mars habitats, even with complicated and energy-intensive filtering and air purification systems. The ubiquitous dust of Mars may, indeed, be one of the big challenges to human development of the planet. This will be especially so if the dust proves to carry the oxidants discovered by *Viking* landers in the soils of Mars.

The Martian canyons—a major group of which is collectively referred to as Valles Marineris—are enormous in size when compared with their terrestrial counterparts. The main canyon is 125 miles (200 km) wide at its widest, and 4.3 miles (7 km) deep, compared with 18.6 miles (30 km) and 1.2 miles (2 km) for Arizona's Grand Canyon. The Martian canyon area in total extends some 2500 miles (4000 km). The canyons are quite rough with many landslides from their walls (figure 2.15), troughs, longitudinal ridges, and considerable amount of rubble on their floors. Some ridges whose tops are below the surrounding terrain are sharply peaked. Those that have tops level with the surrounding terrain are flat topped.

The canyon walls show plain evidence of erosion, with many fluted cliffs, but this

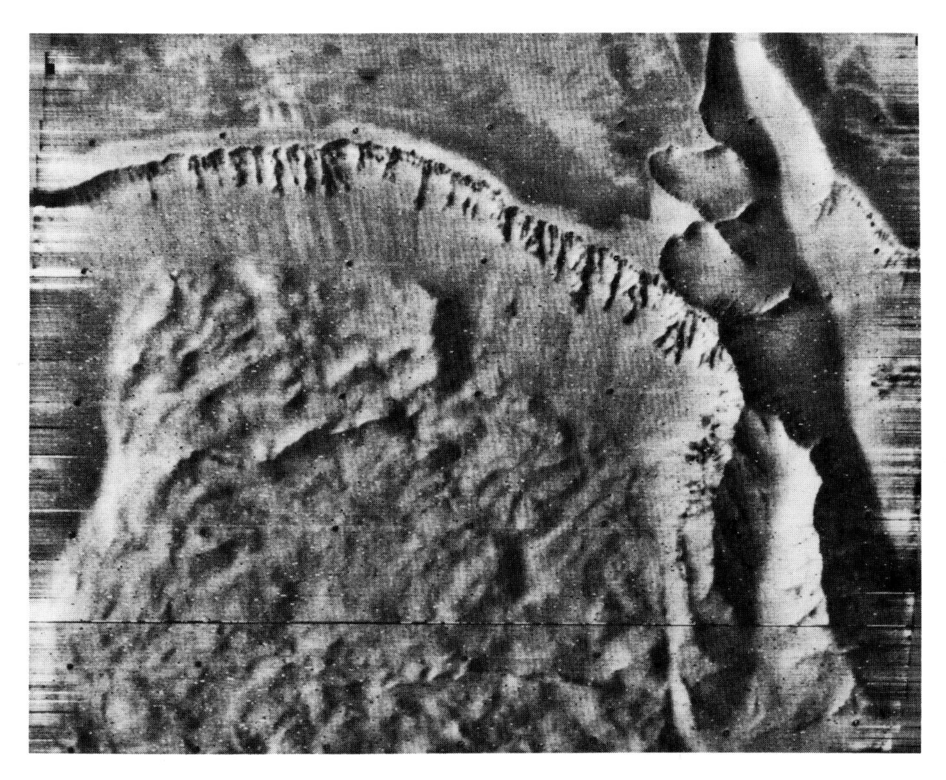

FIGURE 2.15: A massive landslide from the walls of one of the big canyons leaves fluted cliffs in which layering can be seen. There have been suggestions that these layers originated from episodic ash falls from the Tharsis volcanoes. (NASA/JPL)

erosion may not necessarily have resulted from water. However, there are many valleys in the surrounding terrain, deepening as they enter the big canyons. These valleys appear to have been generated by water sapping from the terrain rather than precipitation of rain on the area.

The main canyon of the Valles Marineris group (figure 2.16) consists of three named sections, Ius Chasma, Melas Chasma, and Coprates Chasma. It extends 1500 miles (2400 km) approximately east to west, trending slightly southward, with other canyons such as Tithonium Chasma, Hebes Chasma, Ophir Chasma, and Ghangis Chasma tending to be aligned parallel to it. At the lower end the main canyon divides into two branches, Capri and Eos Chasmas, and here the surrounding terrain is between 0.6 miles (1 km) and 1.2 miles (2 km) above the mean level. The terrain slopes downward to the north, and many eroded channels and broad valleys run from the lower canyon area to the great northern plain of Chryse.

Another large canyon north of the main canyons borders on Lunae Planum as a wide south-north trough which becomes Kasei Vallis through which large quantities of water once flowed into the west side of Chryse Planitia (figure 2.17). A similar separate canyon to the east, Juventae Chasma, also trends south to north and leads to another set of channels draining into Chryse from the south west.

Beyond the canyons to the east there are large areas of what has been called chaotic terrain, also believed to be terrain that was rich in ices and water which broke up and released the trapped liquid, possibly as a result of volcanic heating.

These great canyons of Mars exhibit thick sequences of layered deposits (figure 2.18)

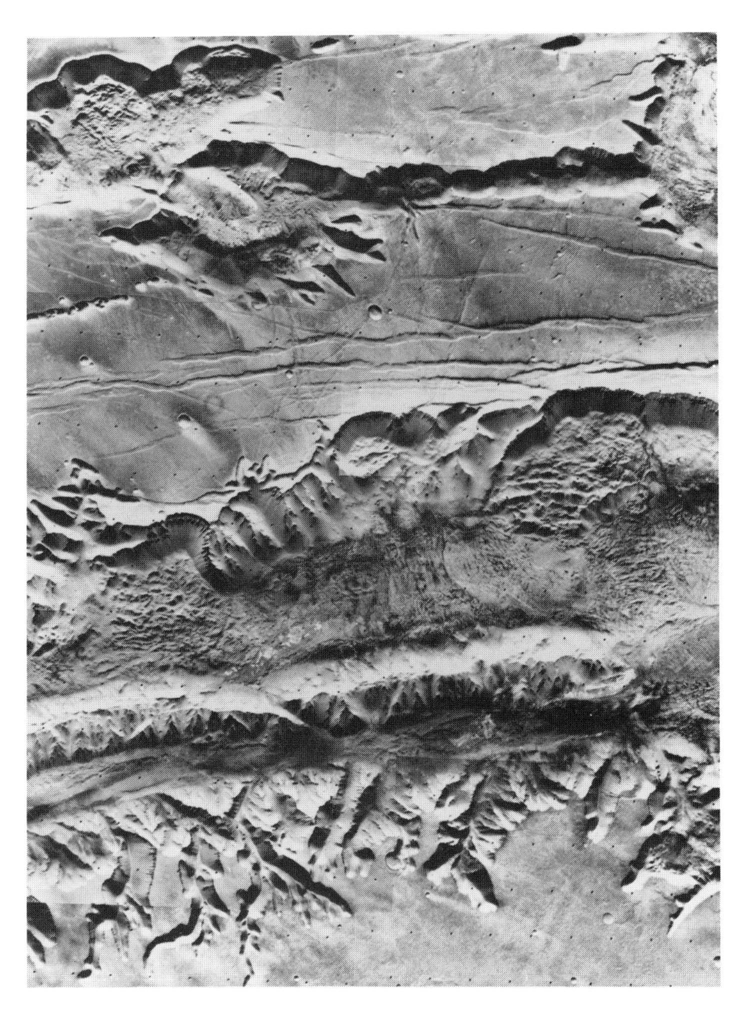

FIGURE 2.16: A more general view of some of the main canyons shows many landslides and intricate details of the canyon floors. (NASA/JPL)

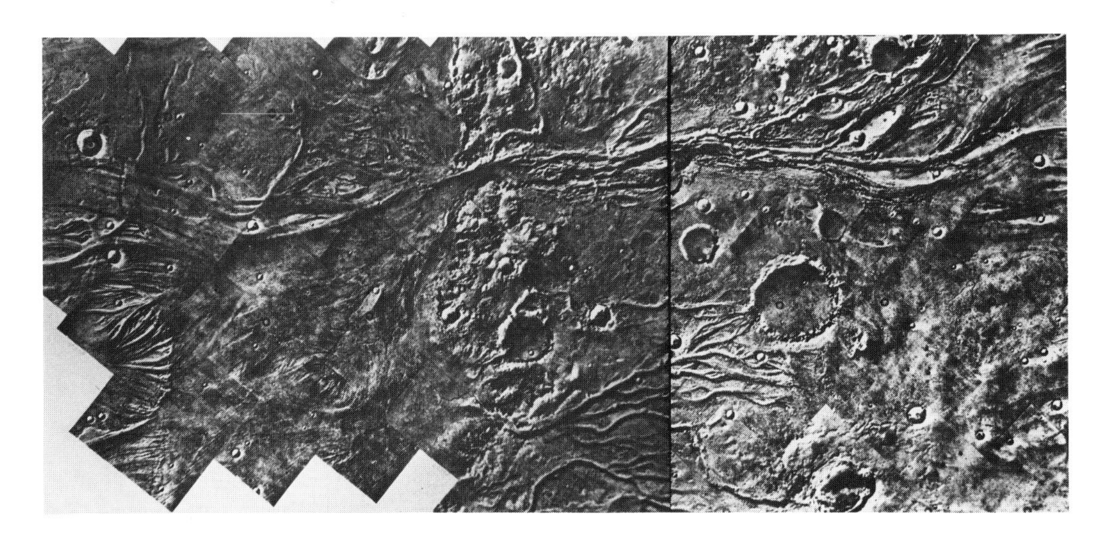

FIGURE 2.17: Large quantities of water apparently flowed from the canyon areas into the west side of Chryse Planitia. This mosaic shows a part of the area molded by the floods. (NASA/JPL)

FIGURE 2.18: A box canyon known as Hebes Chasma is about 95 miles (150 km) long. Layered sediments on the canyon's floor are believed to have been deposited in standing bodies of water and represent the beds of ancient Martian lakes. (NASA-AMES)

which suggest most strongly sediments laid down in bodies of water which at one time flooded these mighty canyons. The layering is mainly horizontal, and at places there are mesas which suggest that surrounding deposits were eroded to leave the isolated table mountains.

The deposits do not resemble sediments from floods, but rather sediments laid down in quiet standing bodies of water over a considerable period of time and with various episodes of sedimentation to produce distinct layers of considerable extent.

A possible scenario is that at the time the canyons were formed by the lifting of the Tharsis bulge there was much liquid water within the regolith of the planet. At the high point of the canyons on the Tharsis bulge is the complex area of Noctis Labyrinthus (figure 2.19) close to where the bulge projects its highest, about 6 miles (10 km) above the mean surface of the planet. This area consists of intersecting closed depressions which to the east become wider, connect one with another, and finally merge into the big canyons.

The faulting of the surface to create the big canyons would be expected to breach underground reservoirs of water. The liquid would flow into the valleys, filling them to considerable depths, possibly to as great as 3 miles (5 kms). As the planet cooled these bodies would have become covered with a permanent ice cap. As ice sublimed at its surface, new ice would form at the base of the ice layer. This process would be expected to continue until all the standing water had been used up, or a subsequent increase in the height of the Tharsis bulge tipped the remaining water out of the canyons so that it flooded onto the northern plains.

FIGURE 2.19: The intricate pattern of faults and valleys is the high part of the Tharsis bulge and is called Noctis Labyrinthus. To the right of the picture the valleys begin to change into the great canyons. (NASA/JPL)

While the water was trapped in the canyons, sediment deposited on top of the ice cover could have worked its way downward into the lakes. Additional sediment could have come from an overload of sediment that broke the ice cover periodically. Other sources of sediments would be collapse of the canyon walls, and volcanic activity on the canyon floors. However, the latter does not appear to have taken place because there is no evidence now of any volcanic structures on the valley floors. There are, however, many examples of landslides from the walls spilling across the canyon floors.

An alternative explanation is that the layering has resulted from episodic ash falls from the Tharsis volcanoes. False color images of the canyon area reveal sheets of color which have been interpreted as resulting from extensive layers of volcanic ash.

Over 50 percent of the surface of Mars consists of ancient cratered terrain which has a dense population of impact craters produced by the heavy bombardment to which all planets appear to have been subjected early in the history of the Solar System. This bombardment is known from the *Apollo* missions to have ended about 3.9 billion years ago on the Moon. However, because of the proximity of Mars to the asteroid belt, the bombardment of Mars may have continued to as recently as 2.9 billion years ago. Other areas of the planet which are mostly vast plains, also have a population of craters younger than those in the heavily cratered terrain. Why the heavily cratered terrain should be concentrated in one hemisphere of Mars is as yet an unsolved mystery of the planet.

Some of the great plains have a very sparse population of craters which may imply that the plains units of Mars are still being formed by lava flows. On close inspection the plains exhibit many features typical of lava flows, including wrinkle ridges (figure

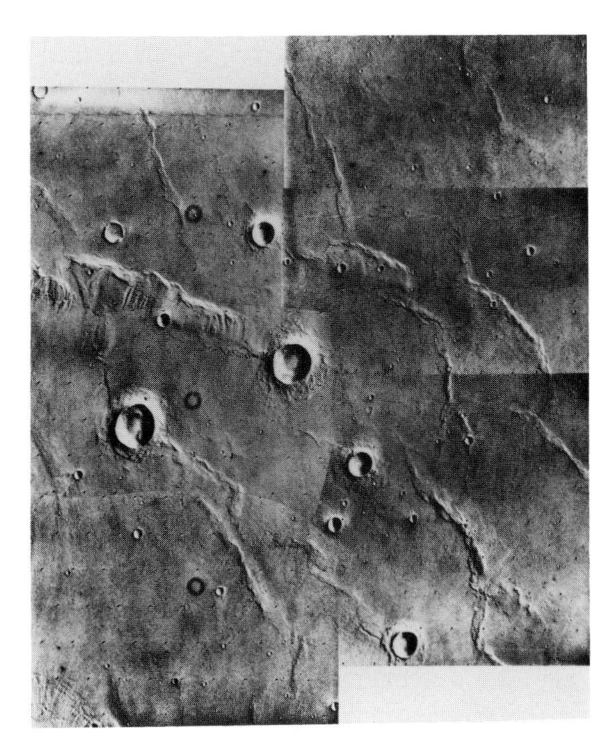

FIGURE 2.20: Areas of Mars are covered by lava plains much like the mare regions of the Moon. On these plains there are fresh impact craters and wrinkle ridges. The picture shows an area on the western part of Chryse Planitia. Grooves in some of the ridges are thought to be the result of water erosion. (NASA/JPL)

2.20) similar to those of the lunar dark areas and flow fronts at the edges of sheets of lava. In other regions the plains are quite different, and this is especially so for the plains surrounding the poles. In those regions there is little visible evidence of lava flows but features that suggest erosion and deposition activities. Around the edges of the polar regions are also vast areas of sand dunes (figure 2.21), and the poles themselves appear to be capped with layered deposits through which some canyons have been eroded.

In the heavily cratered terrain there appear to be two types of craters; large craters whose shapes have been subdued by impact weathering (figure 2.22), and smaller craters which have retained more of their pristine sharpness (figure 2.23). The big craters were probably formed soon after the planet accreted and were subsequently subdued by a heavy bombardment of smaller objects. The fresh-looking craters probably resulted from a continued but declining rate of impact of these smaller bodies. While central peaks are rare in the larger craters, they are more common in the smaller craters.

The unusual feature of these heavily cratered terrains of Mars compared with those of Mercury and the Moon is the abundance of erosional channels (figure 2.24) indicating at least one pluvial period after the craters were formed, but before the formation of the fresh-looking smaller craters. Moreover, these channels are of many types which suggest a long period of fluvial activity because some are quite fresh-looking, while others are barely visible. The period probably began with and continued until after the outgassing of the planet's initial atmosphere. Since the atmosphere at this period was

FIGURE 2.21: The south polar region of Mars is almost free of craters and covered with layered deposits. In this view it is partly covered with the bright cap. (NASA/JPL)

much denser than it is now there was probably also much wind erosion and formation of dune fields as well as erosional deposits from both water and wind. The floors of the larger craters appear extremely smooth, although at high resolution they are seen to be sprinkled with small secondary impact craters similar to what is seen on the smooth-looking floors of lunar craters.

Perhaps the most interesting of the Martian impact craters are, however, the base surge craters which give the impression of having been formed in a volatile rich regolith (figure 2.25). These rampart craters appear even in the equatorial regions while the current theories of location of water ice in the regolith are that it is confined to terrain at latitudes greater than 30 degrees. Generally it has been assumed that water or water ice was the volatile that caused these characteristic craters. If rampart craters must be associated with subsurface volatiles then equatorial regions must have had ice or water relatively close to the surface when the rampart craters were formed.

There is some experimental evidence to support the idea that rampart craters are

FIGURE 2.22: Large craters on Mars have not survived very well from the period of intense bombardment. In some regions of Mars, particularly in that around 30° north latitude, the erosional processes seem to be still active. Rims of larger craters have been truncated, leaving only vague ring-ridges, and smaller craters form subtle, bowl-shaped recesses embossed on pedestals as material outside the craters is preferentially removed. (NASA/JPL)

characteristic of ejectas originating from impacts that were made into a regolith rich in volatiles, but it is difficult to ascertain whether the volatile responsible was water, ice, or frozen carbon dioxide. In fact, some of these craters appear to have two types of ramparts, one that may have originated from water and the other which may have originated from ice at the same site. This could very well be the case since an ice regolith would be expected at some depth to have liquid water below the permafrost layer.

Analysis of gases from an SNC meteorite discovered in Antarctica which is assumed to have originated on Mars, have been interpreted as showing that Mars has much water ice in its regolith, possibly extending hundreds of meters deep globally. Some researchers have identified mud flows, polygonally fractured sedimentary deposits, ridges near the mouths of stream beds, and evidence for interactions between hot lava and ice on Mars.

Rampart craters often have a ridge at the edges of the ejecta flow (figure 2.26), and also are not found with diameters less than a few kilometers. It has been suggested that the craters showing a ridged rampart may be the result of impact into liquid water-rich regolith, while those without were formed by impacts into ice-rich regolith. The concentration of ridged craters in equatorial regions supports this viewpoint, since the regolith would be more likely to have water in the warmer regions at the time when the craters were formed.

Internal radiogenic heating of the planet led to partial differentiation, continued outgassing of the atmosphere, and reworking of large areas of the surface with lava

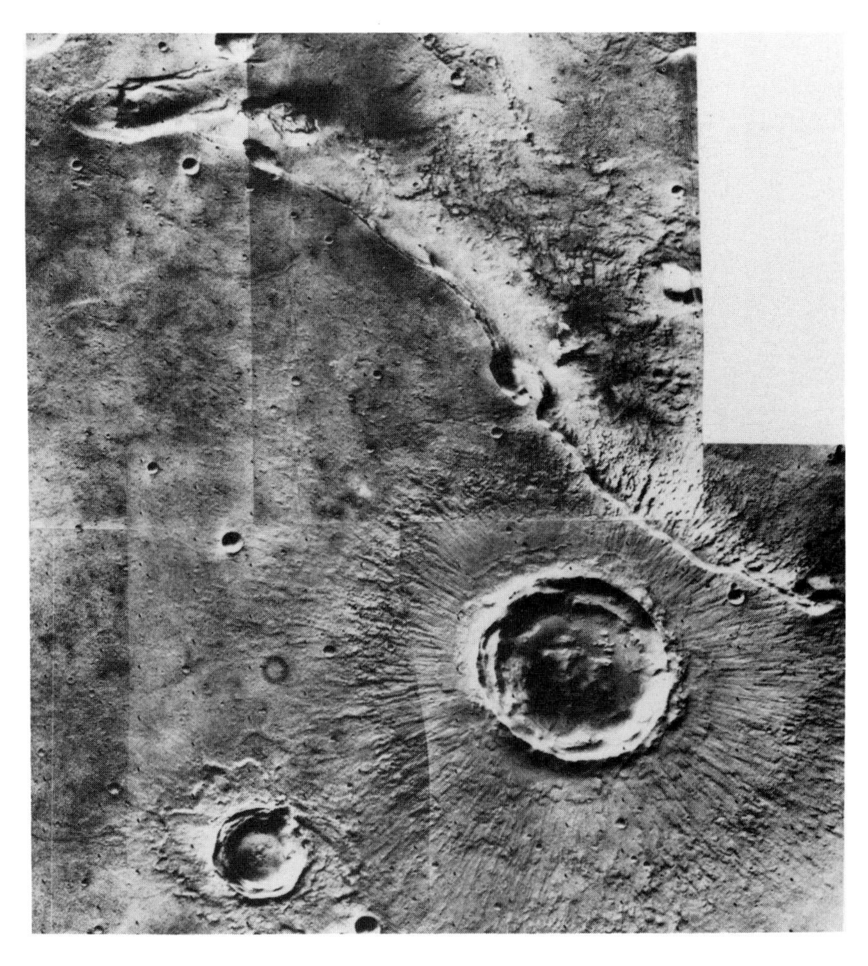

FIGURE 2.23: There are still fresh young craters on Mars. This is about 18 miles (30 km) in diameter and is located on Lunae Planum. It is very much like some lunar craters with terraced walls and hummocky peaks on the dark flat floor of the crater. The dry river valley is Kasei Valley. It is located in what appears to be an area of lava flows. (NASA/JPL)

flows during a period that probably extended from about 3 billion to 2 billion years ago. During this period the dense atmosphere became trapped in the surface rocks. This period merged into the beginning of plate tectonics, the era of the big shield volcanoes and the Tharsis uplift, and the production of the large volcanic plains. In this era, too, the great floods probably swept across the surface as geothermal heat melted ice locked in the regolith from the earlier ages. By a few hundred million years ago the volcanism in Tharsis, the most recent volcanically active area, was dying down, although some probably continues to this day. Figure 2.27 shows images gathered by *Viking* Orbiter on August 30, 1977, which have been interpreted by Leonard Martin of Lowell Observatory, Flagstaff, as a possible present-day eruption from a small crater chain just south of the large volcanic mountain Arsia Mons. In a period of one hour a large disturbance occurred in which local clouds and possibly surface material appeared to have been blown radially outward from the eruption area. The dense eruptive cloud was about 4 miles (6.5 km) wide and reached a height of 3 miles (5 km) above the surface during the period of observation recorded by the Viking images.

Martin also found other evidence of current volcanic activity on Mars. He reported a

FIGURE 2.24: This area west of Argyre has a surface broken by many faults which offset the plains and the impact craters. Also, stream channels with many tributaries may record the effects of water that rained out of the atmosphere early in the history of Mars. The bright areas are part of the extended south polar frost cap. (NASA/JPL)

FIGURE 2.25: Arandas crater is a typical base surge crater. Its diameter is 15 miles (25 km) and it has a central peak. Material around the crater appears to have flowed along the surface rather than being ejected along ballistic trajectories like lunar and terrestrial craters. Experimental craters made in waterlogged ground produce this same effect. Scientists interpret these patterns as being caused by impacts into an ice-rich surface rather than a dry surface like that of the Moon. (NASA/JPL)

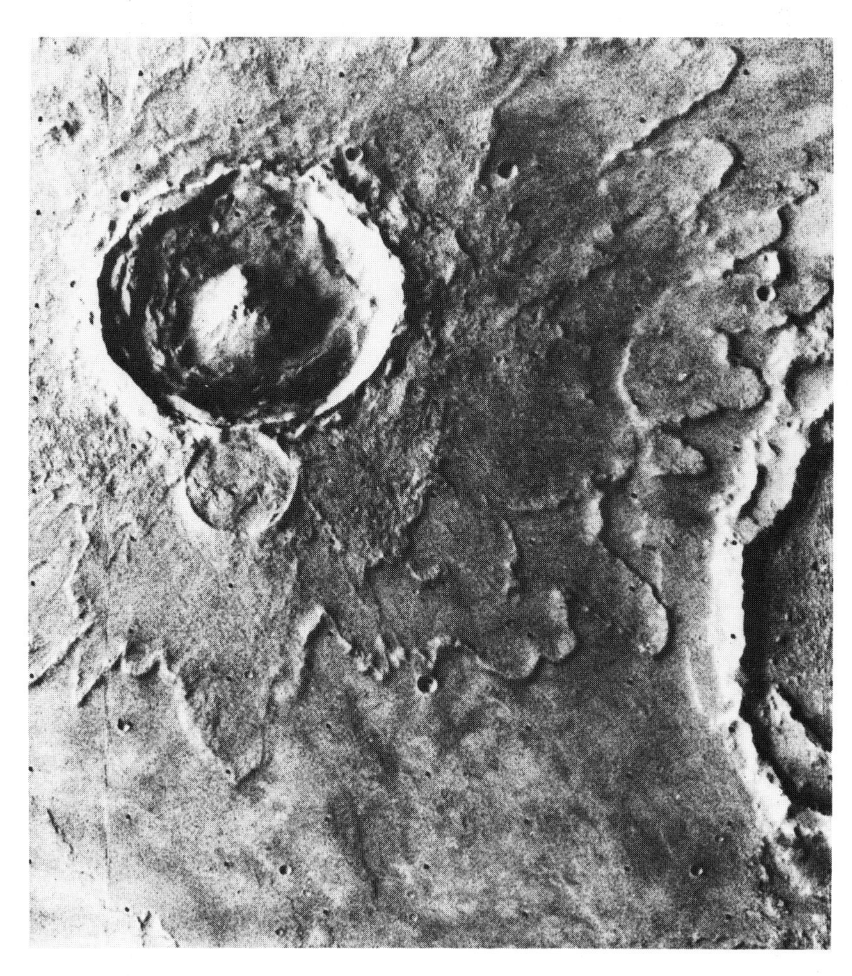

FIGURE 2.26: This 11-mile (18-km) diameter crater, Yuty, is also a base surge type, but the lobate flows are layered. Also, the leading edge of the flow has a ridge in many places. This is similar to the ridges formed by terrestrial avalanches. This crater also has a very pronounced central peak like Arandas. Note how the flows have almost obscured an earlier and smaller crater next to Yuty. (NASA/JPL)

geyser-like eruption north of Solis Lacus, and has suggested that planetwide dust storms might be triggered by volcanic eruptions, citing unusual clouds originating from an isolated peak at the start of a global dust storm during the *Viking* mission to Mars.

Other evidence of recent volcanic activity on Mars was presented by L. Wilson, P. J. Mouginis-Mark, and J. W. Head at the Third International Colloquium on Mars held at the Jet Propulsion Laboratory in 1981. A volcano, Hecates Tholus in Elysium, was shown to have an elliptical area extending from its caldera toward the west for some 20 miles (35 km) in which most surface details are subdued or completely obliterated. The suggestion was that the area represents mantling by material explosively ejected from the volcano, and that nearby radial grooves on the flanks represent pyroclastic flow channels. However, high resolution images of these channels show dendritic patterns which might indicate they are of fluvial origin in which water flow was able to quickly scour materials from earlier ash falls from this volcano. Such channels are not seen in the areas assumed to be covered by material which fell from the air following the more recent explosive erruption. The fact that there are no impact craters on the mantled area attests to its recent occurrence.

FIGURE 2.27: a) Researchers at the Planetary Research Center, Lowell Observatory (led by Dr. Leonard Martin), inspected many *Viking* pictures and deduced that this sequence of three images obtained on August 30, 1977 of an area on the slopes of Arsia Mons showed a cloud (arrow) from a volcanic eruption on Mars, indicating that the planet is still volcanically active. (Lowell Observatory) b) Dr. Martin also identified on the *Viking* images an example of current vulcanism in the Sinai Planum region on the eastern slope of Tharsis. The right picture taken 4.48 seconds after the other on August 1, 1978, shows considerable changes to the cloud shape and shadow. This may imply that the plume was a column of steam from a geyser, or a very small volcanic vent. The shadow is about 1500 feet (500 m) across. (NASA/JPL)

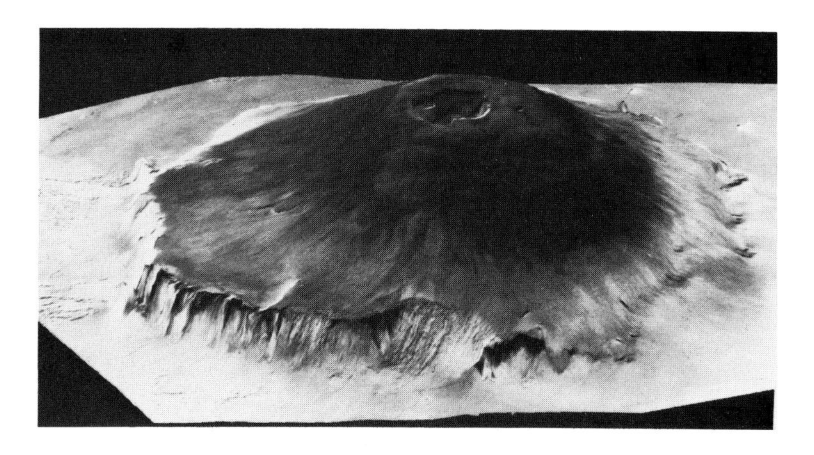

FIGURE 2.28: Olympus Mons is the largest volcano on Mars, and possibly the largest in the Solar System. It towers almost 17 miles (27 km) above the mean surface level of the planet. This computer generated perspective view developed by U.S. Geological Survey's Flagstaff Image Processing Facility provides a good impression of the high cliffs at the edge of the shield and the enormous dome with its complicated caldera. (NASA-Ames-U.S.G.S.)

There are many volcanoes on Mars. While the largest are in Tharsis, Elysium, and Hellas, others are spread around the planet. Some are obviously quite young, others very ancient. They are also of different types ranging from the mighty shield volcanoes such as Olympus Mons (figure 2.28) to small cratered domes and cratered cones. Some of the domes have elongated summit vents and some medium sized volcanoes have arrays of flows similar to the rift zones from volcanoes on the Big Island of Hawaii. There are many volcanic caldera and those on the big Tharsis volcanoes are gigantic compared with their Hawaiian counterparts. The caldera on Olympus Mons (figure 2.29) is 43 miles (70 km) across while the Kilauea caldera (figure 2.30) of Volcanoes National Park, Hawaii, is only about 3 miles (5 km) across.

There is no doubt but that Mars is a planet on which volcanic activity has occurred over most of its history. Michael H. Carr of the U.S. Geological Survey, Menlo Park, has suggested ages for the various volcanic units based on crater counts. He concludes that the Tharsis volcanoes are the youngest, but how young is yet to be determined, possibly only a few million years for some of them. Volcanoes in Elysium are older than those in Tharsis. Older still are volcanic plains of Syrtis Major and Isidis Planitia. And even older are the volcanic units of Hellas and Tyrrhena Patera.

Around the base of Olympus Mons are massive cliffs over which lava has flowed (figure 2.31). These cliffs are 3.7 miles (6 km) high in places, elsewhere they are covered by the lava. Surrounding the base cliffs is a wide concentric area of mysterious ridged terrain which forms an aureole to the mountain. The ridges are roughly parallel, closely spaced, and curve to follow the shape of a lobe (figure 2.32). The outer edge is higher than the surrounding plain and ends in a cliff. The inner edge is lower and becomes covered by younger material from the volcano. What appear to be younger and coarser units in some places overlie older units. Their surfaces have relatively few craters which attest to their being relatively young. Theories abound, but none are conclusive. There have been suggestions that the volcano was surrounded by thick ice sheets which later melted to leave the ridged pattern of debris, that there were explosive eruptions of ash, and that the great mass of the volcano squeezed material up through

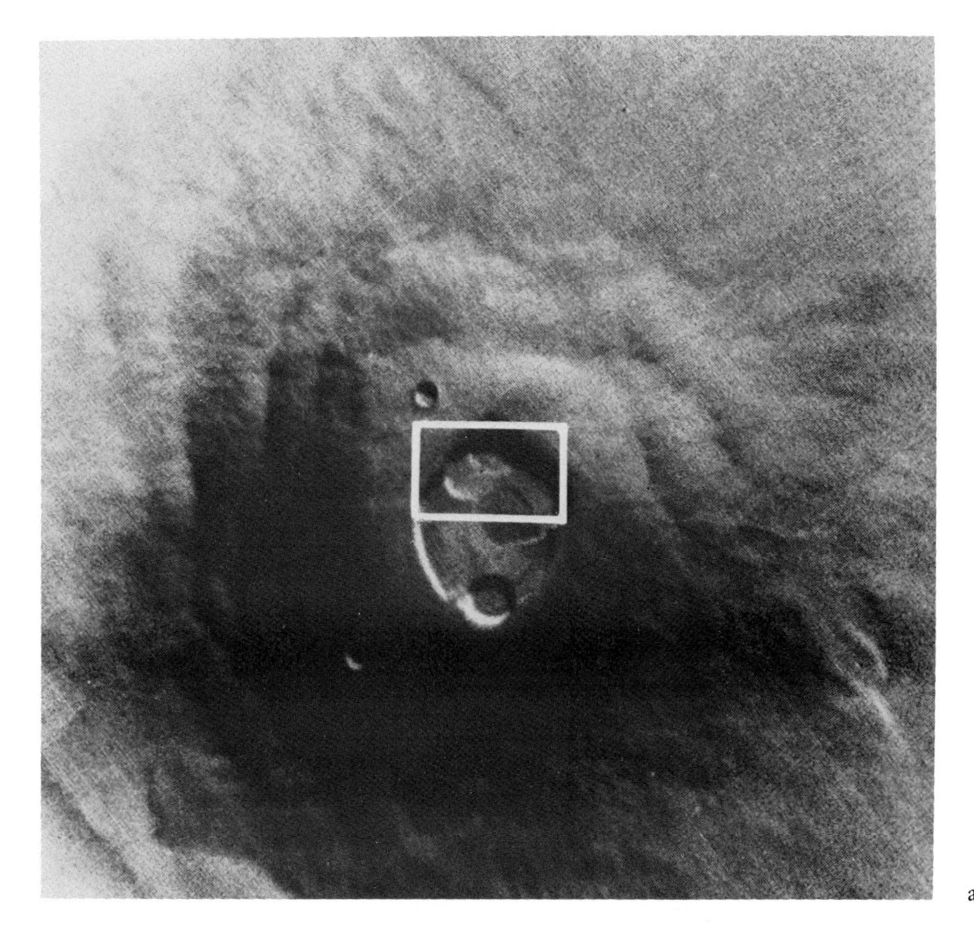
a

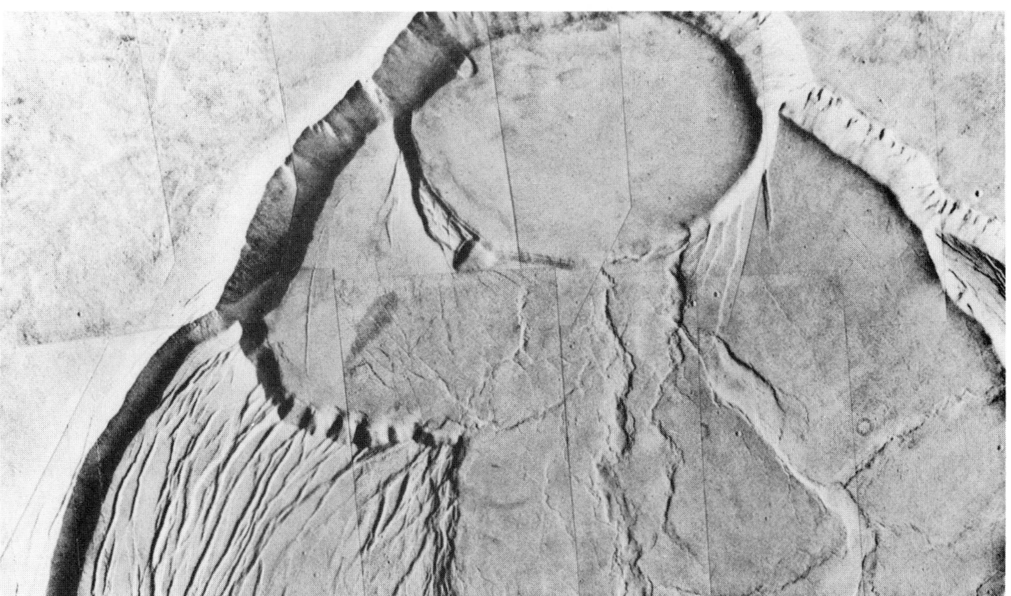
b

FIGURE 2.29: a) The box identifies the caldera area on Olympus Mons which is enlarged in b) to show its structure. This caldera is complex and records a series of eruptions in its varied levels of frozen lava lakes. The summit craters are from 1.5 to 1.7 miles (2.4 to 2.8 km) deep with wall slopes of about 32°. (NASA/JPL)

FIGURE 2.30: Kilauea Caldera in Hawaii has a diameter of only 2.4 miles (4 km). It has several lava lakes floored by lava flows which erupted within the last century. (U.S.G.S)

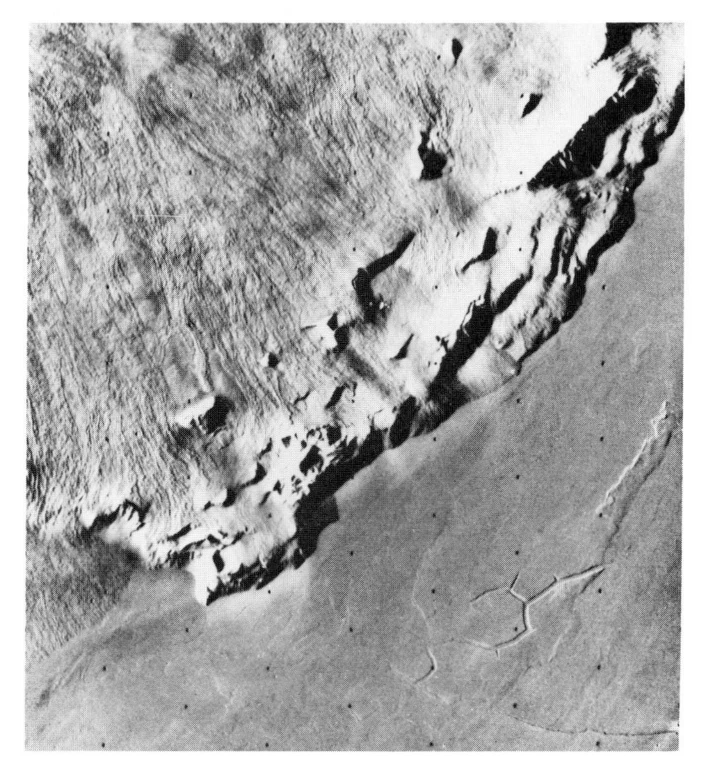

FIGURE 2.31: The base cliffs of Olympus Mons display many faults and lava flows which in some parts virtually eliminate the cliffs. (NASA/JPL)

FIGURE 2.32: A massive landslide from one of the base cliffs of Olympus Mons has produced a lobate fan with a concentric pattern of ridges. Elsewhere in an aureole around the big volcano there are similar ridged patterns on what appear to be lava flows across the surrounding flat surfaces. (NASA/JPL)

fractures to form the aureole. Undoubtedly an answer will be forthcoming only when we can send human explorers into the region.

There are somewhat similar patterned terrains elsewhere on the planet and not associated with any volcano. Examples are in Cydonia (figure 2.33) and Umbra. The origin of these features of regular curved patterns of ridges is also unknown. They seem to be older than the Olympus Mons features because they have some craters superimposed upon them.

The Cydonia area is also of great interest because of some unusual features, areas where great blocks of rocks stand out starkly from surrounding plains (figure 2.34). One of these blocks (figure 2.35) bears an uncanny resemblance to a simian face. This has been explained as a fortuitous chance of lighting. However, there have also been suggestions based on another image at a slightly different angle of sunlight that the shape is quite real and is related in several ways to nearby features on a geometrical basis. There have been suggestions that the "face" on Mars is evidence of extraterrestrial intelligence at some earlier epoch laying down a purposeful pattern on the planet's surface. Whether or not this shape is fortuitous or designed or is another example of

FIGURE 2.33: Peculiar geometric markings so regular that they appear almost artificial are seen in this area of Cydonia. The contoured markings are in a shallow basin possibly formed by wind erosion. The markings are about one half mile from crest to crest of the low ridges. (NASA/JPL)

anthropomorphic Mars remains to be seen when the area is again surveyed, and in more detail, by one of the upcoming missions to Mars. Certainly the area of Cydonia has some unusual formations, but so does most of Mars. Whether or not we read into these formations evidence of intelligent life on ancient Mars, or visits from extraterrestrials even, may be very important to some people on Earth. Generally, however, this current controversy goes to show that Mars was, is, and will continue to be an intriguing planet, a planet that stimulates the imagination and challenges us to explore it in detail to satisfy the insatiable curiosity of the human spirit.

Many areas of Mars show mysteriously patterned ground (figure 2.36) such as stippled areas within smooth plains, mottled areas around craters, knobby terrain, fretted terrain, and polygonally fractured terrain, to name a few. Some terrains are extremely complex and difficult of geological explanation. The polar terrains of Mars offer perhaps the greatest challenge to our understanding. During spring in the planet's northern hemisphere the seasonal carbon dioxide cap sublimes into the atmosphere and its edges retreat toward the pole. By the beginning of summer it has vanished completely to reveal the underlying cap of frozen water. This ice cap is a source of water

FIGURE 2.34: Cydonia is also an area of sculptured features and eroded mesa-like landforms. The huge rock formation near the center resembles a human head and was originally interpreted as being formed by shadows giving the illusion of eyes, nose, and mouth. The feature is 1 mile (1.5 km) across, and the sun angle is approximately 20°. The speckled appearance of this image is due to bit errors emphasized by enlargement of the picture which was taken by *Viking 1* Orbiter on July 25, 1976, from a range of 1162 miles (1873 km). (NASA/JPL)

vapor for the atmosphere. Some studies have suggested that there may be a net loss of water and a gradual transference of water from the northern cap to the southern cap under current conditions. It could be that this transfer is part of the process leading to layering at the poles.

However, other models have suggested that the water is returned in full to the northern cap, together with dust, during global dust storms, which could also lead to layering of polar deposits. This question may be resolved when another orbiting spacecraft can make year-long observations of the planet specifically to solve the question. The planned *Mars Observer* mission should be able to do this.

Surrounding the Martian poles are areas of sand dunes (figure 2.37). In the north polar region the dune fields appear to be concentrated close to steep scarps. P. C. Thomas of Cornell University found that this is, indeed, so, and that the material of the dunes appears to be moving away from the scarps; that is, away from the polar regions. He also concluded that the material of the polar dunes is much the same as that in dunes found in craters and canyons elsewhere on the planet (figure 2.38).

There is little doubt that the polar layered deposits (figure 2.39) record variations of climatic conditions on Mars over several hundred million years, as suggested by Michael H. Carr. The age of a few million years at least is confirmed by the presence of several impact craters in the south polar cap that was initially thought to be devoid of such features. The presence of these craters also implies that layering occurs at a very low rate of deposition. Otherwise the craters would have been obliterated. If, however, the layering process has been in process for many hundreds of millions of years and is somehow tied into the cyclic effects of obliquity changes, how is it that the deposits are

a

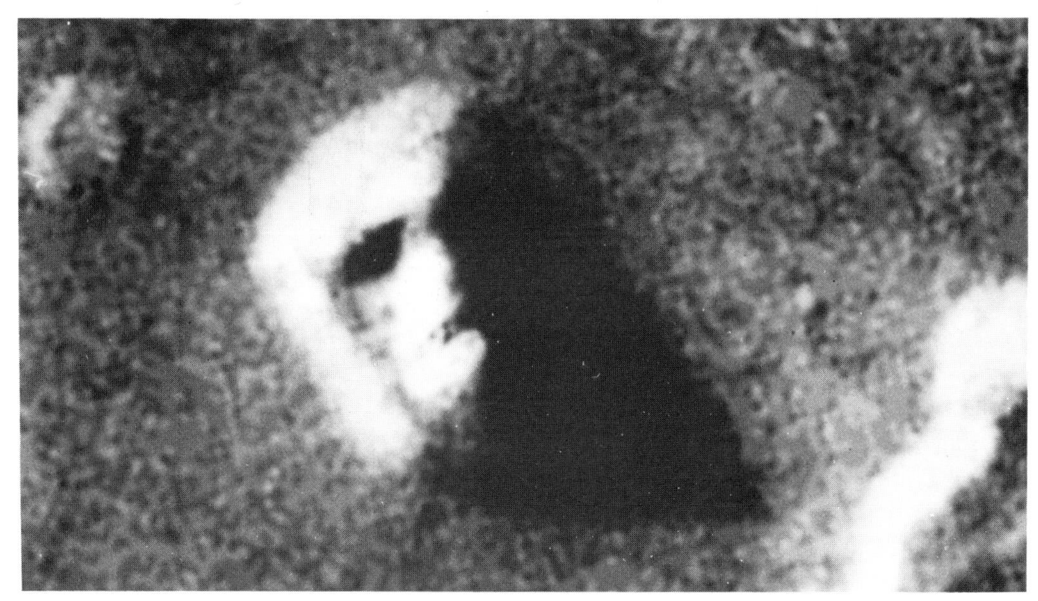

b

FIGURE 2.35: The images of the area have been enhanced by Dr. Mark J. Carlotto working from the raw data supplied by NASA. a) On this reprocessed image of the general area the face appears very simian, and nearby structures to its left look extremely mysterious and interesting. Some features have the sharp angular form of typical pyramids, and a two-sided part of a triangle just left of the picture's center piques the imagination. b) The face itself when enlarged shows interesting details in the halftones. Dr. Carlotto has also processed another image in this same area taken at a slightly different sun angle, and the resemblance to a face is still very strong. However, some researchers have claimed that in another image taken at a very different sun angle the resemblance is lost. These pictures were made available by The Mars Project, a Berkeley-based organization which recommends that this area of the Red Planet should receive much more detailed scrutiny at the highest possible resolution during future missions to Mars to resolve what has now become a somewhat heated though very interesting controversy.

limited in thickness above the surrounding plains? One explanation is that the deepest layers of ice and dust are under sufficient pressure and high enough temperature for the ice to change state to liquid water which flows underground away from the poles. Thus, as new layers are added at the surface, layers beneath the surface disappear and the polar deposits remain about the same thickness.

The removal of ice can be accounted for in this way. The dust also can be removed from the polar regions by erosion and redistributed elsewhere on the planet. These polar regions must be given a high priority for exploration in future missions because they hold important information about the geological and climatological history of the planet. Seismic exploration of the polar regions should determine whether or not basal melting is taking place below the layers of ice and dust.

The causes of ice ages on Earth are not certain. It is believed that changes in the orbit of the Earth or inclinations of the terrestrial axis to the orbit can affect climate on the planet on a global scale. Other speculations are that ice ages result from changes to the radiation emitted by the Sun. If we can determine that changes on Mars were synchronized with changes on Earth, then it would be more likely that changes in the output from the Sun is the cause. Instruments developed for use in Antarctica to probe into the ice sheet and obtain long-term records of solar activity might be used with penetrators or landers on Mars to do the same on the Martian ice caps.

Researchers at the University of Arizona compared high resolution photographs of six Martian volcanoes with the Hawaiian volcanoes. On the Hawaiian volcanoes valleys which are known to be caused by water erosion had many similar characteristics to the valleys surrounding the volcanoes on Mars. Moreover, the researchers concluded that contrary to the generally accepted theory that water flowed on Mars only very early in the planet's history, Alba Patera and its valleys indicate that water flowed on Mars much later. Additionally it seems that the valleys on Alba are very similar to valleys of terrestrial volcanoes that we know were formed by surface water runoff, that is from rain. This might indicate that there were times in the past of Mars when rains fell upon the Martian surface.

While valleys on the slopes of volcanoes can be started by runoff, as the valleys deepen they can encounter underground water trapped behind dykes. When such dykes are breached by runoff erosion, floods can be released to rapidly gouge out larger valleys.

Other researchers from U.S. Geological Survey, Flagstaff, have studied the channel system in the Cerberus region at the edge of Elysium Planitia (figure 2.40), and concluded that this is the youngest valley system on Mars. This valley system stretches for about 2000 miles (3200 km) from Elysium into Amazonis. Materials from the channels are slightly darker than the plains on which they have been deposited. Tear-drop-shaped streamlines indicate the direction of flow. The young age of these flow deposits is inferred from the paucity of craters upon them. But how recently in geological time the water flows occurred is difficult to determine.

Nearly all the major outflow channels (figure 2.41) which are now generally believed to have carried floods of water on Mars, ultimately lead to the northern plains of the planet. These northern plains of Mars have interesting areas of polygonally fractured ground (figure 2.42); for example, in Utopia/Aetheria and Acidalia. Another area shows chaotic terrain in the low-lying Magaritifer Sinus between Tharsis and cratered high-lands around Deucalionis Regio. Comparison of these landforms with terrestrial ana-logs suggest that the plains once were covered with large bodies of water in which

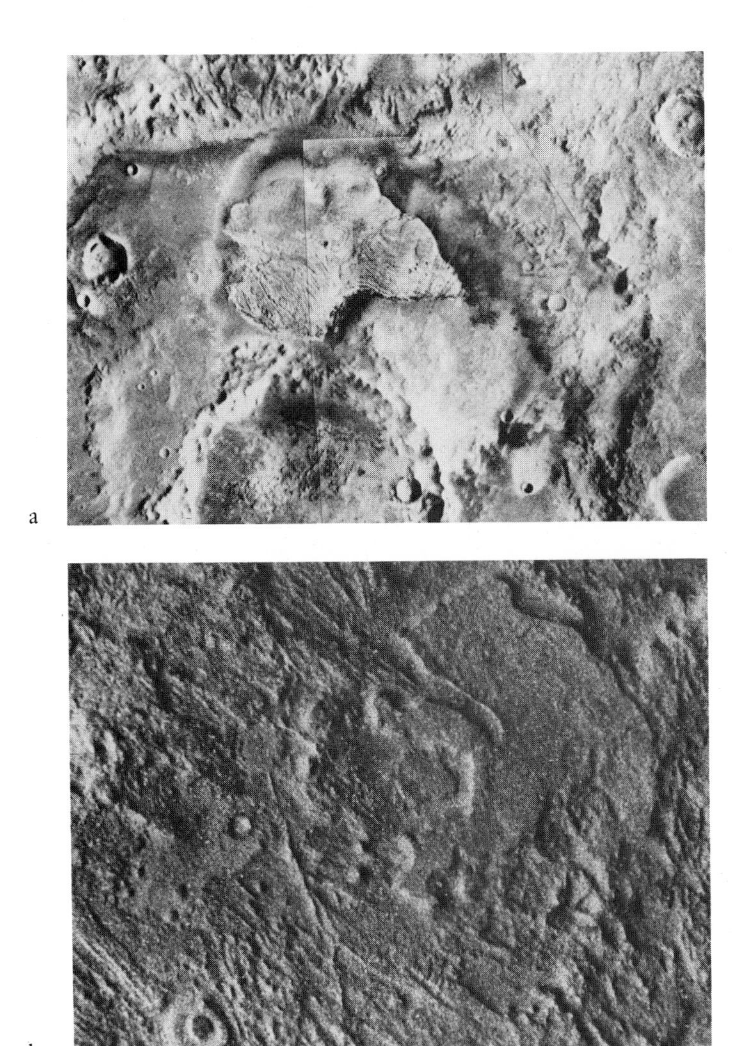

FIGURE 2.36: Many areas of Mars exhibit extremely interesting and somewhat mysterious features that are difficult to explain.

a) This is an area within Becquerel, a 93-mile (150-km) diameter crater located at 20° N., 30° W. An anomalous leaf-shaped deposit is wedged between the walls of Becquerel and the rim of another, smaller crater inside Becquerel. There are grooves and terraces forming closed loops, somewhat similar to the layered deposits at the poles. Why this layered terrain should be here is not understood. One suggestion is that it is another piece of evidence of cycles of major climatic changes on Mars. (NASA/JPL)

b) An example of etched terrain in which layers of surface have been eroded and transported elsewhere leaving tablelands and pedestals. (NASA/JPL)

c) Patterned ground in Cydonia. At the bottom right are the blocks referred to earlier in connection with the "face." This area merges into the great swathe of fractured terrain across the bottom of the mosaic, which is covered to the north by lobate flows from the large impact crater at top left. (NASA/JPL)

d) Another strange feature, nicknamed "White Rock" was first seen in images obtained by *Mariner 9.* It was recorded again in greater detail in this *Viking* image obtained September 1, 1978. The white area measures 8.5 by 11 miles (14 by 18 km). It is located near the equator in a 58-mile (93-km) diameter crater. It is located too close to the equator to be ice. It may be a wind eroded remnant of a layered deposit from a period of different climate. (NASA/JPL)

e) Yet another strange terrain is this area in Novus Mons at 66° S., 325° W. A smooth but striated area stands at a higher elevation than the surrounding irregular terrain and extends about 45 miles (75 km). Cliffs are sharply defined, and the surrounding areas show a variety of albedos. (NASA/JPL)

c

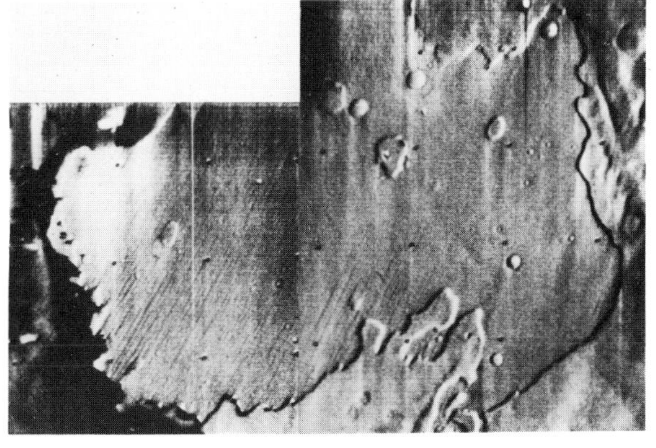

d, left
e, right

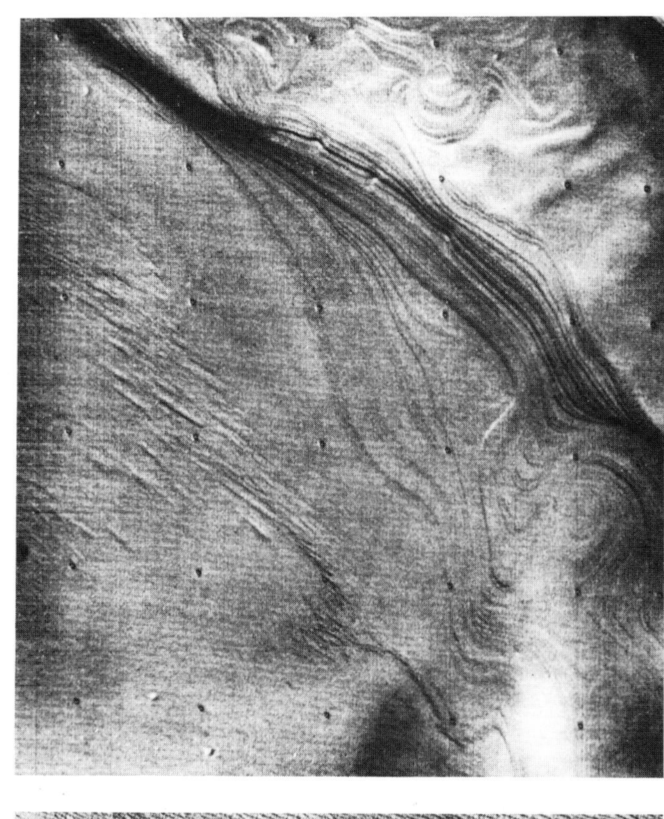

a

b

FIGURE 2.37: Surrounding the poles are areas of dunes: a) shows the typical layered terrain from which the material of the dunes is thought to originate by weathering; b) is a high-resolution image of one of the dune fields extending from a valley in the polar deposits. (NASA/JPL)

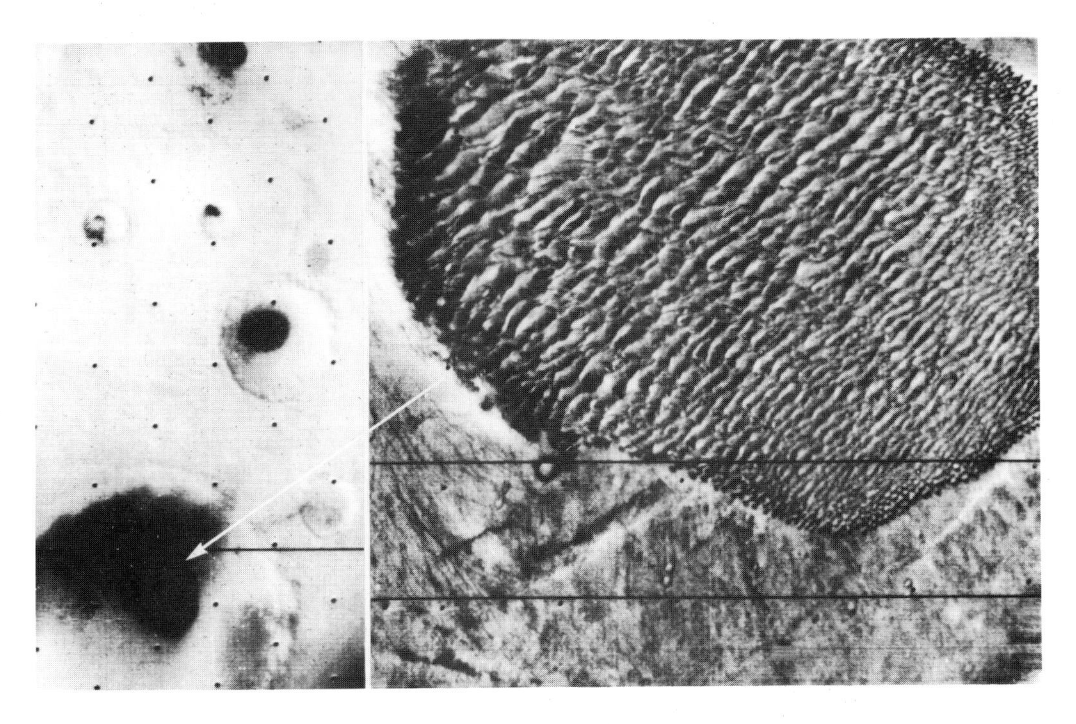

FIGURE 2.38: The material of the polar dune fields appears to be similar to that of other dunes elsewhere on the planet such as those shown here on the floor of a crater in Hellespontus. The crater is 93 miles (150 km) in diameter. The dune field appears as a dark spot in the wide-angle photograph of the crater at the left. In the narrow-angle view of the crater floor, at right, the dune field is shown to consist of numerous long dunes spaced about 1 mile apart with smaller dunes at the field's margins. They were formed by consistent strong winds blowing from the southwest. (NASA/JPL)

sediments carried into these 'oceans' were deposited. The sediments gave rise to the polygonal pattern of fractures which arose when the remaining water froze. The chaotic terrains of Mars may have originated when ice-rich sediments were disrupted by the ice disintegrating because of heat released by volcanic activity.

To deposit sediments over large areas rather than as overlapping fans would require a deep body of water unless the fans were subsequently molded by changes in level of the surface to produce the uniform deposits of today. This is unlikely.

As the planet cooled ice sheets may have formed and these ice sheets might have flowed like those of Earth's Antarctic continent. This viewpoint is supported by sinuous ridges on the low side of highlands similar to those around Antarctica. These observations suggest that water was still present on the Martian surface beyond the first billion years, implying that the climate must have been warmer, or the water must have been extremely saline.

An outstanding question that will probably have to wait for a new landing on the Red Planet is the nature of the oxidant in the Martian soil. Also we need to know if there is water elsewhere on Mars to destroy this oxidant or is the oxidant ubiquitous on the surface. If there are oxidant-free sites would biological molecules be able to exist in the soil without destruction? We would like to know more about the composition of the Martian soil. Does it contain nitrogen, phosphorous, electrolytes, and water? Many gaps in our knowledge of conditions on the Red Planet still exist. They must be filled before we can start a further search for life there.

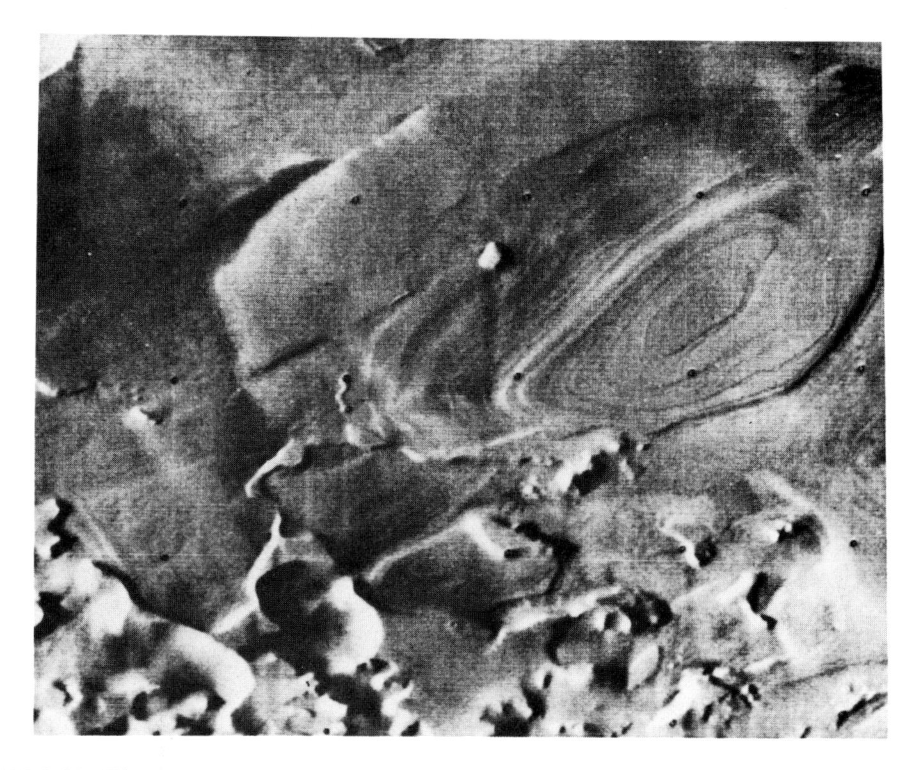

FIGURE 2.39: There now seems little doubt that the layered deposits at the polar regions record climatic changes on Mars. This oval-shaped layered tableland is near the south pole of Mars. Emerging from beneath these deposits an older unlayered deposit is deeply etched with jagged pits and grooves. The area covered by this *Mariner 9* image is 29 by 37 miles (47 by 60 km). (NASA/JPL)

On the desert world of Mars today the average atmospheric pressure at the surface is only about 7 mbar, and although the atmosphere is predominently carbon dioxide its greenhouse effect is small because of the low density. The greenhouse effect on a planet is a result of selective absorption by carbon dioxide and water vapor which are opaque to infrared radiations. The surface of Mars is warmed a mere 5 K to produce an average temperature of 212 K, or −59° C. At the equator during the early afternoon on Mars the temperature can rise to above the freezing point of water, but the low pressure causes the water ice to literally boil off into vapor and in doing so it is kept below the melting point; the water changes from ice to gas without a liquid phase. The only places liquid water might appear sporadically on the surface would be at the bottom of deep canyons or craters where the atmospheric pressure is higher. Unfortunately, as mentioned previously, none of these locations is close enough to the equator for temperatures to rise above freezing.

Temperature at the equator of Mars can reach 16° C, but even summer at the north pole is only −68° C. No place on Earth has conditions as cold as those on Mars. Mars is also extremely dry. Because of the low atmospheric pressure on Mars, ice does not melt into water as the temperature exceeds 0° C. Instead it passes straight into vapor. The triple point of water at which water can exist as ice, a liquid, and a vapor is at a pressure of 6.3 mbar, but the pressure on Mars is generally less than that except in the lower parts of the surface. For example, at the deepest part of the Hellas basin, the

FIGURE 2.40: Elysium Planitia is another large volcanic area on Mars on which two major volcanoes are distinct to the left upper center of this mosaic. Cerberus is the dark elongated area below the volcanoes, and the crater, Mesogaea, with the streaks discussed in figure 2.14 is near the bottom right of the picture. A number of fresh-looking young channels are visible to the left of the two volcanoes. These were probably cut by water originating when ice was melted by volcanic heat. (NASA/JPL)

pressure is 13 mbar. But these low lying regions are all at latitudes where the temperature never rises above freezing. In fact, water migrates in the atmosphere to cold regions of the planet where it condenses on the surface.

The northern ice cap of Mars consists of water ice and is permanent, though it does shrink during the northern summer. By contrast the southern ice cap generally disappears almost completely in the southern summer. This is because the ellipticity of the orbit of Mars brings the planet 27 million miles (43.45 million km) closer to the Sun at southern summer than at northern summer.

There were many questions remaining from the *Viking* experiments. The *Vikings* did not carry instruments capable of detecting the presence of water on or under the surface. The instruments could not measure hydrogen, oxygen, nitrogen or carbon in the soil. The composition of the soil was inferred from examination of spectra. The intensity of ultraviolet radiation at the surface was not measured. Visible evidence of life was generally not thought to be shown by the lander images, but there are claims that greenish patches on the rocks and the soil near *Viking 1* changed over a period and may be evidence of microbial life.

Viking did not determine whether or not there is or was life on Mars. Evidence of water flowing, large amounts of ice in the regolith and polar caps, existence of salts on the surface, are favorable to life. However, against life being on Mars today is the evidence for strong oxidants being present in the soil, the lack of organics in the soil, and the low probability of liquid water being present at any location on the planet.

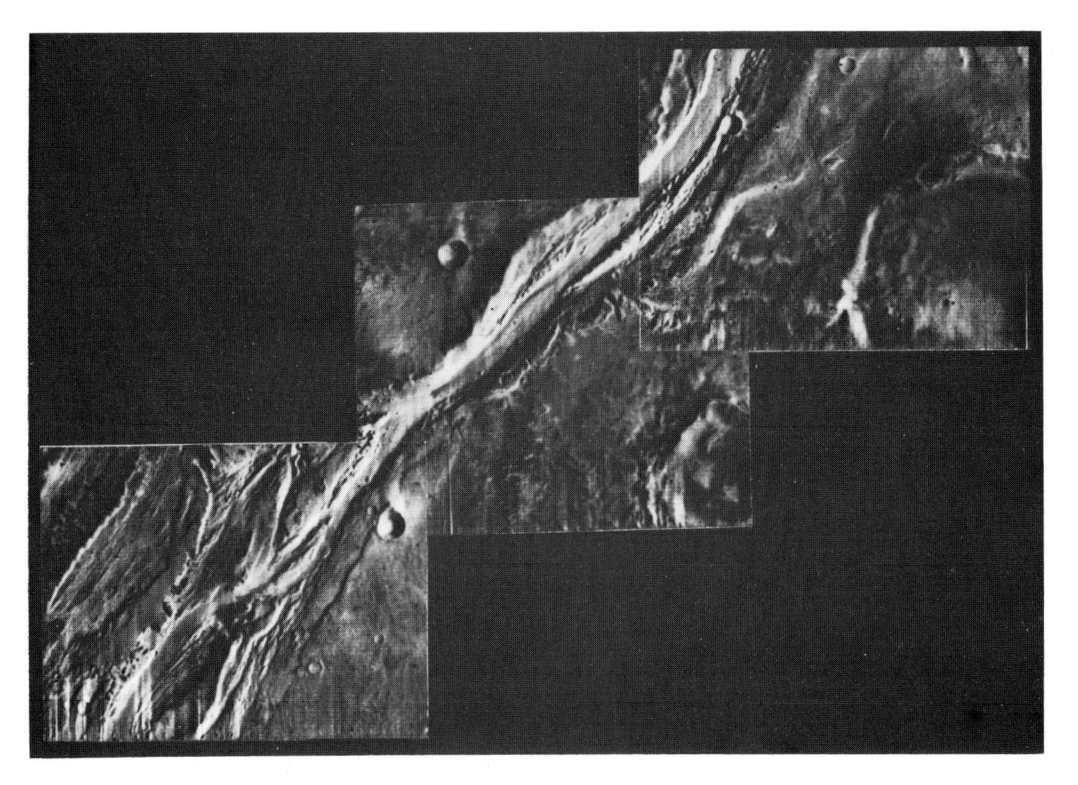

FIGURE 2.41: A small segment of a channel thought to have been formed by running water in Mars' past is located just north of the equator between Amazonis and Memnomia. The flow was from lower left to upper right, and the segment is about 46 miles (75 km) long. (NASA/JPL)

Moreover, despite the fact that carbon is the fourth most abundant element in the universe and complex carbon compounds are widely distributed throughout the galaxy and are an essential part of life as we know it, very little carbon, and no carbon compounds were found on Mars.

The *Viking* orbiter showed that the surface of Mars is not homogeneous but varies considerably from place to place. The dust, however, is most likely ubiquitous because of the planet-wide dust storms, and this dust may contain the strong oxidants detected at the two landing sites. However, at various depths below the surface, the physical and chemical characteristics of the regolith may vary considerably from place to place.

There is support for the theory that Mars once possessed a dense atmosphere of carbon dioxide and had a wet, warm climate early in the history of the planet. If the surface pressure of the carbon dioxide atmosphere sometime in the past was within the range of 1 to 5 bars, solar heat would have been trapped by the greenhouse effect and the surface temperature would have been high enough for liquid water not to freeze.

A geochemical cycle of carbon dioxide was modeled by J. B. Pollack of NASA-Ames Research Center and others who concluded that such a dense atmosphere could have existed early in the history of Mars and could have persisted for a billion years if the carbon dioxide in the atmosphere was replenished at a rate sufficient to balance the extraction of carbon dioxide by weathering. A most likely source would be volcanic activity thermally decomposing carbonate rocks. Nevertheless, the amount of carbon dioxide in the atmosphere would have inevitably decreased as the volcanoes became

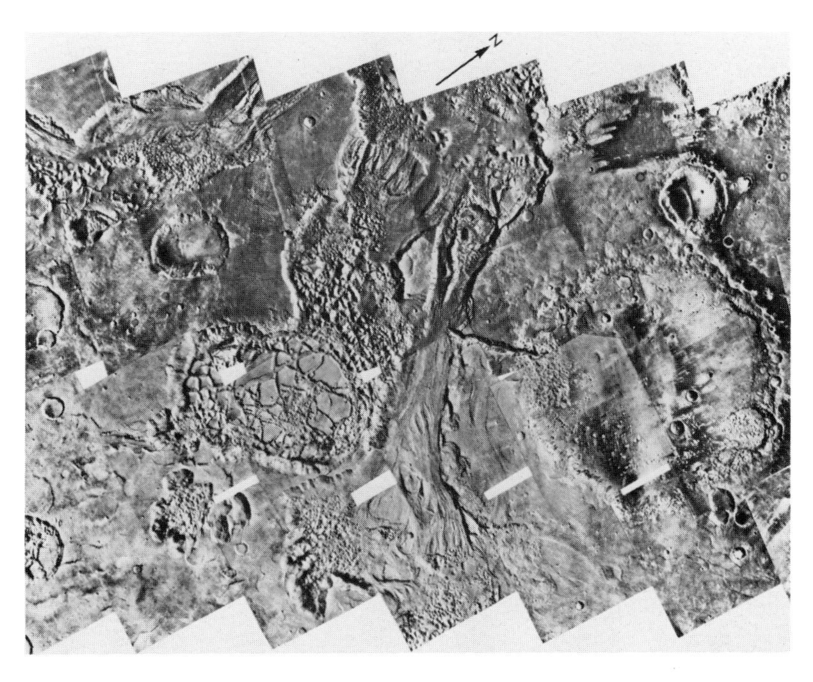

FIGURE 2.42: This area near the eastern end of the Valles Marineris is dominated by the large and very ancient crater Galilaei, at the right of the mosaic, in which there are many smaller and younger craters on its floor. Also, several large collapse features up to half the size of the crater, at left center, are associated with outwash channels. A possible explanation for the fractured and jumbled terrain associated with these features is a subterranean melting of ice and the subsequent drainage of water leading to collapse of the unsupported surface. (NASA/JPL)

less active, the climate cooled, and the water became frozen into the regolith. This theory can be tested in future missions to Mars by searching for the presence of carbonate rocks in the planet's crust.

While the channels discovered on Mars (figure 2.43) appear to indicate that water once flowed in large quantities on the Martian surface, some of these flow features may have been produced by water flowing beneath a cap of ice rather than freely on the surface. However, it is difficult to accept this process as being valid for the intricate valley networks which are also prominent on Mars, especially on areas in the southern hemisphere of the planet. These networks are difficult to explain except by flows of liquid water unrestrained by an ice layer. There are some suggestions that flows were the result of rainfall, but the evidence is not conclusive. They appear to be more likely the result of sapping processes.

There are several large impact basins evident on Mars. The more ancient of these basins, together with the secondary craters produced by debris thrown out by the impact, are more heavily eroded than more recent basins and their secondary craters. This is especially true of features that appear to be dated earlier than the impact basin of Argyre. After the impact that created the Argyre basin, the erosion and the liquid flows appear to have been curtailed. It is thought that the denser atmosphere and the wet period of Mars thus predated the formation of Argyre. Indeed, the Argyre impact might have contributed to the loss of the ancient denser atmosphere.

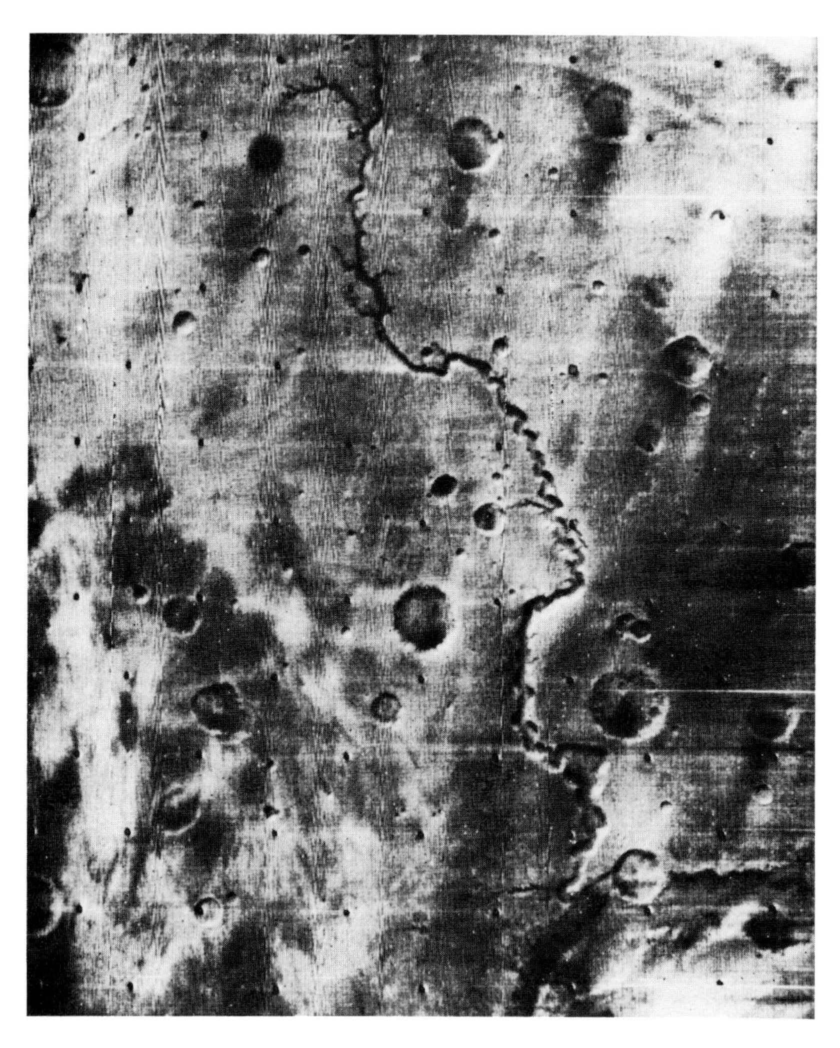

FIGURE 2.43: A dendritic channel on Mars in the heavily cratered terrain indicates that water once flowed freely and copiously on the surface. (NASA/JPL)

The amount of carbon dioxide needed to raise the surface temperature of Mars above the freezing point of water is consistent with estimates of the amount of carbon dioxide that would be initially available on the planet considering its size and likely thermal history.

Carbon dioxide is removed relatively rapidly from the Earth's atmosphere by the presence of water which gives rise to chemical weathering. In fact, some estimates state that all the carbon dioxide would be removed from Earth's atmosphere in only 10,000 years if the carbon dioxide were not replaced by other natural processes.

On Earth the silicate rocks of the continents are weathered and carbonate rocks are deposited on the floors of the oceans. When dissolved in water, carbon dioxide forms a weak acid which erodes the continental rocks. Eventually the metallic ions formed are precipitated as carbonates in the oceans. For a similar process on Mars, the Red Planet would have had to have running water, exposed silicate rocks, and water-filled basins.

To replace the carbon dioxide thus removed from its atmosphere, Mars would also have had to have a source of carbon dioxide to maintain a wet, warm, and denser atmosphere. One obvious source is outgassing through volcanoes of carbon dioxide originally incorporated in the crust when the planet formed. Another is a recycling of carbon dioxide when deposited carbonate rocks were heated by volcanic action.

Early in the history of Mars, the heat flux from the interior was probably several times higher than is estimated for today, namely, about one quarter to one third that of Earth. Calculations show that this would not be sufficient to replenish the carbon dioxide unless Mars was assumed to have a much larger original content of carbon dioxide than is generally accepted. However, if carbon dioxide was also recycled when deposited carbonates were decomposed by hot lavas, the dense atmosphere may have been maintained for a billion years.

However, if Mars had large bodies of water similar to Earth's oceans, much carbon dioxide would be dissolved in the water and this would require an even greater amount of carbon dioxide to be generally available. If, however, the Martian oceans were much shallower than the terrestrial oceans, the planet could have still maintained a carbon dioxide atmosphere dense enough for a greenhouse effect to keep the bodies of water liquid on Mars.

Michael H. Carr of U.S. Geological Survey, Menlo Park, estimated that Mars may have had sufficient water to provide a planetwide ocean of 1600 ft (500 m) deep. This would also imply that sufficient carbon dioxide would be outgassed for an atmosphere of 3 bars surface pressure. This, too, would fit in with the theory of an early wet and warm Mars on which life might have developed in the Martian oceans.

An expedition to the Red Planet may not find extant life, but if it finds carbonate rocks there will be a reason to search more diligently for extinct life as discussed elsewhere in this book.

In looking at the evolution of the Martian climate as suggested by the Pollack model of an early Mars, we see enough carbon dioxide outgassed early in the history of the planet to raise the surface temperature above the melting point of water ice and to sustain a greenhouse effect by recycling carbon dioxide through volcanic action. The heat flux from the interior of Mars would be sufficient to maintain these conditions for about 1 billion years.

This early Mars would have had relatively warm temperatures over most of the planet for the whole Martian year. Later, as carbon dioxide removal exceeded its replacement, the benign conditions would be restricted to regions closer to the equator and at lower elevations; finally, only at times when the planet was close to perihelion. If, as suggested, Mars experiences cyclical climatic changes due to precession and its elliptical orbit, the polar regions, too, might have retained a climate benign to life at epochs and during summer. Life forms might have adapted to these conditions, lying dormant when the water froze, and becoming active in the increasingly shorter periods of benign conditions. Ultimately the conditions would have deteriorated to the harsh frozen desert climate we see today. Was life able to adapt?

While Mars became drier and colder, there were possibly a few periods of warmer climate when spurts of volcanism occurred. At such times large amounts of carbon dioxide discharged into the atmosphere from volcanoes might have temporarily raised atmospheric pressure, caused a short-lived greenhouse warming, and permitted liquid water to appear again on the surface for a while. Did life forms become active again during such periods? The Viking experiments showed no evidence of living things on Mars today at the two lander sites. They also showed a very reactive surface that would be inimical to life. If these surface materials are carried around the planet by the dust storms, the soil could be inimical to biological molecules everywhere on Mars. Survival of living things on the Martian surface today does not seem likely, except possibly beneath the surfaces of rocks. But could life have retreated from the surface to establish underground habitats nutured by volcanic heat? On Earth, for example, microorganisms such as bacteria, algae, fungi, and protozoans, have been discovered living at a depth of 1000 feet beneath the surface.

Answers to such questions may only be possible when we determine that Mars did, indeed, have a period early in its history when life developed on the Red Planet, and subsequent periods when that life became able to adapt gradually to the changing climate and able to evolve into more suitable habitats.

3

A QUESTION OF LIFE

Since seasonal changes were first observed on Mars early in the telescopic era, scientists and laymen have speculated on the possibilities of the planet being the abode of life. These speculations suggested that Mars was a planet with an older and more advanced civilization that that of Earth. Initial space age exploration of Mars negated these ideas and trended toward accepting Mars as a desolate and lifeless planet. More recently the trend has reversed. The discoveries that organic molecules are common in space and that life was present on Earth at least 3.5 billion years ago, coupled with a better understanding of the formation and early evolution of planets, have stimulated much scientific interest in a return to Mars. There is growing support for a more detailed examination of the Red Planet even to the extent of maintaining a permanent human presence on Mars similar to what we have in Antarctica. The important questions of how life evolved on Earth, and did it also start on other worlds, seem intrinsically dependent on understanding Mars. In recent years considerable new knowledge about life has been accumulated. It is important to review this new viewpoint before discussing in more detail how we can further explore Mars to search for extinct or extant life.

Radio astronomy has reported finding more than 60 organic molecules in space. Indeed, molecules comprising carbon and hydrogen appear to be quite stable in space. Molecular ices of water, ammonia, and carbon monoxide together with alcohols, nitrates, aldehydes, esters, and carboxylic acids have been found in material from which stars are forming. This suggests that complex biogenic compounds are being produced even during the early phases of evolution of stars. It appears that organic molecules are present throughout space, presumably many are formed there in interstellar clouds, and also they were in the solar nebula from which planets, satellites, and comets were formed. The building block precursors of all the biomolecules known on Earth are found in space. Small biopolymers were on the primitive Earth, and presumably also on Mars and other worlds.

In the Solar System comets (figure 3.1) are known to contain many prebiotic precursors; they are veritable storehouses of organics. Meteorites also have been found to contain amino acids, the building blocks of living things. Meteorites striking Earth as it formed billions of years ago may have supplied the chemical building materials for the first cell membranes, according to University of California at Davis zoologist, David W. Deamer. Extensive structural and chemical analyses of pieces of the Murchison meteorite, which fell to Earth near Murchison, Victoria, on September 28, 1969, have revealed the presence of lipid-like compounds similar to those that make up the membranes of common cells. In fact, one compound isolated by researchers forms spontaneous membranes around what appears to be droplets of nonmembranous material. It is speculated that winds and waves in ancient terrestrial oceans swept these membrane-forming substances onto beaches and tidal pools where the compounds accumulated. These membranes enclosed the genes and proteins of the first cells that were able to replicate and become living things.

Earlier, in 1970, Dr. Cyril A. Ponnamperuma, then with NASA's Ames Research Center, had examined the Murchison meteorite for presence of amino acids. He found that the meteorite did contain at least five of the twenty amino acids normally found in living cells. Most important, these amino acids in the meteorite were equally distributed between two types, one that predominates on Earth and another that doesn't. Had the Murchison amino acids resulted from terrestrial contamination, the terrestrial dominant amino acid would have been expected to dominate the meteorite. Because it did not, the amino acids must have been within the meteorite when it fell to Earth; they must have been synthesized somewhere else in the universe.

Estimates have been made of how much cometary material has been accumulated on the Earth. During the first two billion years of Earth's existence as a planet the estimates are that it received many times more organic molecules than exist in all terrestrial life now. There seems no reason to doubt that Mars was supplied with plenty of prebiotic molecules at the time the planet formed and during subsequent influx of comets and meteorites. Today there appears to be large areas of photochemical oxidation on Mars, with the result that no organic components can exist on the surface soil of the planet. However, there is currently speculation, particularly in the Soviet Union, that anaerobic bacteria, deriving energy solely from volcanic heat, may have survived on Mars and may still be present below the surface.

A current belief is that the three planets, Venus, Earth, and Mars were very much alike soon after their formation from planetesimals about four billion years ago. Their crusts contained similar materials, their atmospheres contained similar gases, and their average temperatures were much the same and sufficiently high for liquid water to exist on the surface.

Some planetologists contend that the reason the three planets subsequently evolved quite differently was because of differences in the way they were able to process carbon dioxide between atmosphere and crust. Carbon dioxide plays an important role in creating the climatic conditions on a planet. The gas allows solar radiation to reach and

FIGURE 3.1: Comets carry many prebiotic precursors from which life may originate as cometary material fell onto planets. Comet Bennett photographed by the author, April 1970.

heat the surface, but does not allow the infrared radiation from the surface to escape back into space. The carbon dioxide in a planet's atmosphere thus acts like the glass of a greenhouse and increases the planet's temperature over that which an airless body would have at the same distance from the Sun.

On Earth the processes are such that if the surface of the planet should cool because of reduced incoming radiation from the Sun, more carbon dioxide enters into the atmosphere than is removed from it, and the greenhouse effect raises the temperature back again. If the planet warms, an excess of carbon dioxide leaves the atmosphere and the greenhouse effect is reduced so that the planet cools down. By contrast Venus could not compensate for surface heating by removing carbon dioxide, and a runaway greenhouse led to today's extremely hot and dry planet with a dense carbon dioxide atmosphere. Mars, on the other hand, was not able to release enough carbon dioxide back into the atmosphere and it became a cold dry planet with a tenuous carbon dioxide atmosphere. Some recent theoretical studies suggest that large impacts were able to blow off much of the early atmosphere of Mars as a result of the planet's small size and weak gravity. Escape velocity from Mars is only 3.1 miles per second (5.0 km/sec) compared with 6.93 miles per second (11.18 km/sec) for the Earth.

Astrophysicists believe that the Sun emitted less radiation at the time when the Solar System formed and for a period afterward. Since that time the luminosity of the Sun has gradually increased to about 1.5 times its early value. A major question is why the Earth was not originally a frozen world since we know that the Earth was warm enough to have oceans as long as 3.5 billion years ago. The question can be answered if the atmosphere of Earth changed. A lower albedo resulting from fewer clouds would trap more solar heat, and an atmosphere with more carbon dioxide would have trapped even more solar heat despite the weaker solar radiation. In fact, the Earth could very well have been warmer than it is today.

Carbon dioxide is removed from Earth's atmosphere when it becomes dissolved in rainwater to form carbonic acid which, in turn, weathers rocks to produce carbonates that are carried into the oceans. Living organisms in the oceans consume the carbonates and incorporate them into shells which later collect on the floors of the oceans. However, in the course of time because of the active plate tectonics of the Earth, the carbonates in the rocks of the ocean floor are subducted beneath the continents by sea floor spreading. Subjected to higher temperatures and pressures, the carbonates become incorporated into silicate rocks with the release of the carbon dioxide which is vented through volcanoes back into the atmosphere.

The whole process may be acting on the Earth like a planetary thermostat. Should the average temperature of the Earth fall, there is less evaporation from the oceans and less carbon dioxide absorbed by rain, and accordingly less is removed from the atmosphere. The regeneration from the floor of the oceans continues undiminished, however, so that carbon dioxide continues to enter the atmosphere. The result is more atmospheric carbon dioxide, a greater greenhouse effect, and restoration of the original temperature. The converse is that if the planet warms up, more rain falls, more carbon dioxide is removed. As before, the rate of outgassing continues unchanged. The result is that the greenhouse effect is reduced and temperature is again restored to normal. How effective this thermostat may be over the long term is not yet really proved.

Also the biosphere may be acting as a fine control on the planetary thermostat. Plants remove carbon dioxide from the atmosphere by photosynthesis. When they die organic material is deposited in sediments which later are acted upon by tectonic processes or are oxidized to release carbon dioxide back into the atmosphere. Changes in temperature affect the proliferation of plants with correcting effects on the amount of carbon dioxide that they remove from the atmosphere while the return to the atmosphere remains fairly constant.

By contrast with carbon dioxide, water vapor in the atmosphere, which also provides a greenhouse effect, enhances changes in temperature. A higher temperature puts more water vapor into the atmosphere and increases the greenhouse effect to raise the temperature further. A lower temperature reduces the amount of water vapor and pushes the temperature still lower.

On Venus a surplus of deuterium may imply that the planet's ancient ocean was lost into space; hydrogen escapes more easily into space than the heavier deuterium atom. Because Venus is closer to the Sun, it received more heat than did Earth, and the greenhouse effect of its primitive, predominently carbon dioxide, atmosphere changed the water of the oceans into vapor which enhanced the greenhouse effect. The water molecules were broken down by solar ultraviolet with the consequence that hydrogen escaped into space and oxygen reacted with lava flows and became lost to the atmosphere. The lack of oceans on Venus prevents plate tectonics; the vulcanism is vertical. The absence of plate tectonics is inferred from inspection of radar maps of the planet. These radar maps show hundreds of volcanoes and basaltic lava flows. Current vulcanism is inferred from the fact that the gravity anomalies on Venus correspond to topographic features. On Earth such anomalies disappear within a million years, so we know that Venus must still be volcanically active for the anomalies to persist.

What about Mars? The evidence appears to be that Mars was warm enough in the past to have had liquid water on its surface. This presumably was because it possessed a strong greenhouse effect from a denser carbon dioxide atmosphere. Carbon dioxide was probably removed from the Martian atmosphere the same way as on Earth, although, as mentioned earlier, impacts may have been a major cause of atmospheric depletion. The big difference, however, is that because Mars is a smaller planet it did not possess sufficient internal heat energy to develop plate tectonics even though it possessed much water. After some major volcanic episodes during which some carbon dioxide was probably recycled to keep a moderate climate for possibly a billion years after the planet's formation, the planet slowly died. Volcanic activity was considerably reduced until today it is probably mostly confined to the Tharsis volcanoes. Consequently carbon dioxide could not be recycled back into the atmosphere but remained locked in the rocks. The greenhouse effect diminished until the planet froze. A possible confirmation of this hypothesis is the discovery by a NASA scientist that a special type of meteorite found on Earth contains calcium carbonate. These SNC meteorites, mentioned in the previous chapter in connection with water in the Martian regolith, are believed to have originated from a major impact on Mars which ejected fragments of the Martian crust into space later to fall to Earth.

By international agreement meteorites are named after a nearby geographical feature, such as a town or a prominant natural landmark close to where they are found. The SNC meteorites are named after the localities where the first discovered had fallen; Shergotty in India, Nakhla in Egypt, and Chassigny in France. They are achondrites and are made up of igneous rocks. They do not have the small spherical silicate bodies known as chondrules which are common in another and large class known as chondritic meteorites. Some of the SNC meteorites are basaltic and appear to have originated from lava flows. These SNC meteorites are assumed to originate from Mars because age-dating of their materials suggests they were formed only 1.3 billion years ago, much less than the 4.5 billion years age of most meteorites. They were unlikely to be from the Moon or from asteroids because those bodies cooled several billion years ago at least. Some meteorites recovered from Antarctica have properties similar to those of rocks from the lunar highlands; quite different from the SNC meteorites. The only likely origin of the SNC meteorites is Earth, which is still volcanic, Mars on which we see the extensive volcanic areas, or Venus which radar images and gravity anomalies suggest has also been and still is volcanically active.

It is unlikely that the SNC meteorites originated from the Earth because they are known to have been in space for a long time. They also have a different oxygen isotopic composition from terrestrial rocks. At a conference on Lunar and Planetary Science at the Johnson Spaceflight Center, Houston, in 1983, Donald Bogard of NASA introduced measurements of noble gases trapped in the SNC meteorites to show that their proportions were similar to those in the Martian atmosphere as determined by Viking experiments. The planet of their origin appears more likely to be Mars than Venus. If this is true, we have in the SNC meteorites samples of Martian crust to analyze even before we obtain samples returned by spacecraft.

Elemental and isotopic abundances of atmospheric gases trapped in an SNC meteorite are very similar to those measured by *Viking* in the Martian atmosphere. Based on such correlation, other abundances that were not established by *Viking* have been assumed. For example, Martian and terrestrial elemental abundances are remarkably related and if the relationship can be extended to water, Mars should be well endowed with water.

Trace amounts of carbonates are also present in SNC meteorites, and isotopic ratios suggest that these originated from atmospheric carbon dioxide which, in turn, suggests that the carbonates were fixed by life processes. If so, life may have evolved on Mars to a stage at which it could fix atmospheric carbon dioxide in carbonate rocks.

Some evidence has, also, been presented that suggests the elemental abundances of the noble gases on Mars and Earth were similar. If this is so, and current abundances are explained by later depletion of noble gases on Mars, the initial water abundance on the planet could have amounted to sufficient water for a planetwide ocean some 12 to 18 miles (20 to 30 km) deep. However, the estimates based on the volatile elements in SNC meteorites suggest an ocean of only 330 feet (100 meters). This is an enormous range of possibilities. Between these extremes is an estimate made on the basis of the material eroded from the Martian flow channels on the assumption that at least as much water must have flowed as the volume of eroded material. On that basis the planet could have had an initial water budget sufficient to provide oceans of 1600 feet (500 meters) deep planetwide. However, in all cases, the estimates do not necessarily mean that the water would be on the surface as oceans. Much would be trapped in the rocks, in minerals, in the regolith, and as permafrost beneath the surface of the planet.

The consensus appears to be, however, that Mars initially had enough water for there to be standing bodies of water on the surface early in the history of the planet. Today there might still be the equivalent of a global ocean 0.6 miles (1 km) deep buried in the regolith which is thought to extend to depths of up to 6.2 miles (10 km) or so beneath the surface. Of this water, about half may be as ice but the rest, at lower depths, must be in a liquid state.

Some planetologists suggest that there is evidence of current glaciation on Mars. This is seen in Elysium and at the edges of some mountain ranges and appears to be the result of slowly flowing glaciers from which water is ablating. Other scientists suggest, however, that the landforms are not slowly flowing glaciers but mixtures of ice and rocks slowly flowing across the surface to produce characteristic lobate shapes.

Recapitulating the geological features of Mars, we can then look again at the planet from the standpoint of there being extant or extinct life. As with other planets, Mars is not a perfect sphere, it is slightly pear-shaped in that its center of mass is offset into the northern hemisphere while its center of volume is in the southern hemisphere. The departure from a sphere is slight and, of course, not apparent in a telescopic view. The surface of the southern hemisphere is higher in general than that of the northern hemisphere. Gravity measurements made so far by observing the perturbations to spacecraft that have orbited the planet imply that the planet has a thick, rigid crust. This is confirmed by the size of the big volcanoes of the Tharsis region. The mantle appears to be cooler and thicker than that of the Earth, and the lithosphere which includes the crust of the planet is thought to be about 30 to 110 miles (50 to 175 km) thick.

The soil of Mars as sampled by *Viking* contains abundant oxygen, probably as much as 40 percent. In fact, it appears that the red color of Mars results from the presence of iron oxides in the soil. The elemental composition of the soil resembles that of clay; it includes silicon, iron, aluminum, magnesium, calcium, sulfur and traces of chlorine, titanium, and potassium.

At the surface the atmosphere consists 95 percent of carbon dioxide, 2.7 percent of nitrogen, 1.6 percent of argon, with traces of oxygen, carbon monoxide, neon, krypton, xenon, and ozone; not very hospitable to terrestrial life forms. The surface pressure is very low, about 7 mbar but it varies quite considerably over a Martian year as carbon dioxide freezes out at the polar caps in winter and sublimes back to the atmosphere during summer. A significant part of the atmosphere condenses out over the pole located in the winter hemisphere.

Relative humidity is high in the lower atmosphere and peaks at night. So atmospheric water is available. There are morning fogs, polar hoods, and other water clouds (figure

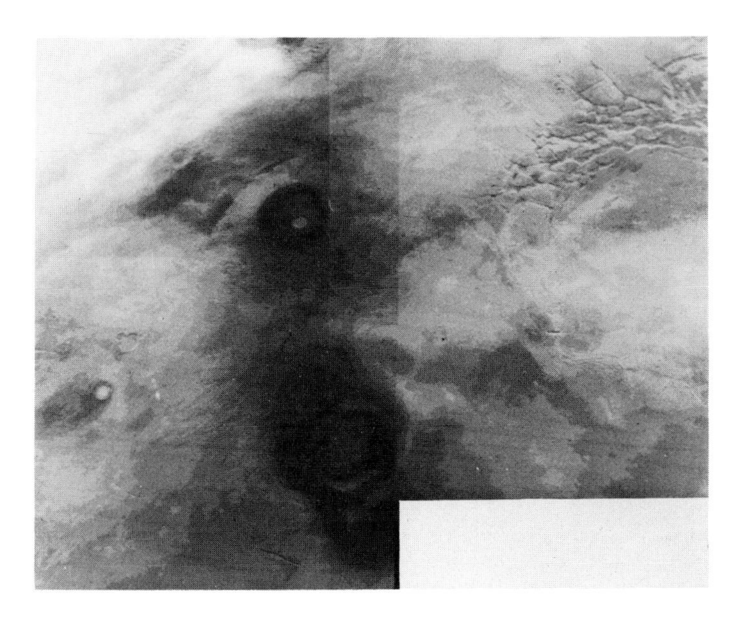

FIGURE 3.2: An oblique view across the Tharsis Ridge taken July 13, 1980 shows the dark summits of the volcanoes Arsia Mons and Pavonis Mons. Two smaller volcanoes, Biblis Mons and Ulysses Patra, are on the left side of the image. It was late northern summer on Mars and the area is dominated by clouds at this season. The valleys at the lower right of the image are filled with haze. (NASA/JPL)

3.2). Where moist air moves up on encountering the big volcanoes, it forms huge clouds in the local afternoon, clouds that have been observed from Earth. Mars has a cloud coverage on the average of about 5 percent compared with Earth's coverage of nearly 50 percent. While the white Martian clouds are of water, frequent yellow clouds are dust clouds. At times the yellow clouds can obscure much of the surface and they sometimes spread planetwide (figure 3.3). No terrestrial weather condition compares with these global dust storms of Mars.

During most of the Martian year the winds are less than 11 miles per hour (18 km/hr), but when the planet is near perihelion the summertime atmospheric temperature rises and winds of 80 miles per hour (130 km/hr) can be whipped up. Local dust storms are fed until they merge into global storms, and the dust absorbs solar heat to keep the "pot boiling." The dust may remain in the atmosphere for a third of the year and by blocking sunlight cause climatic changes. As the surface temperature falls, the winds die down, and the dust settles out of the atmosphere. These planetwide dust storms do not, however, occur every year.

Mars has a residual polar cap of frozen carbon dioxide and water in the south and of frozen water in the north, so there is plentiful frozen water in the polar regions. Why the two poles differ so much is not certain, but the south cap is brighter than the north cap and presumably the high albedo prevents absorption of sufficient solar heat to sublime all the carbon dioxide there during the southern hemisphere summer.

Wind streaks on the surface (figure 3.4) reveal the direction of flow of the planet's wind systems. These streaks form most frequently between late summer and early fall in the southern hemisphere when the dust is settling out of the atmosphere. Some of the streaks are the result of dust being removed from the landscape, others are from deposition of dust. Certain areas show a consistent pattern each year. The seasonal removal of dust from some large areas, such as Syrtis Major, and its deposit on other

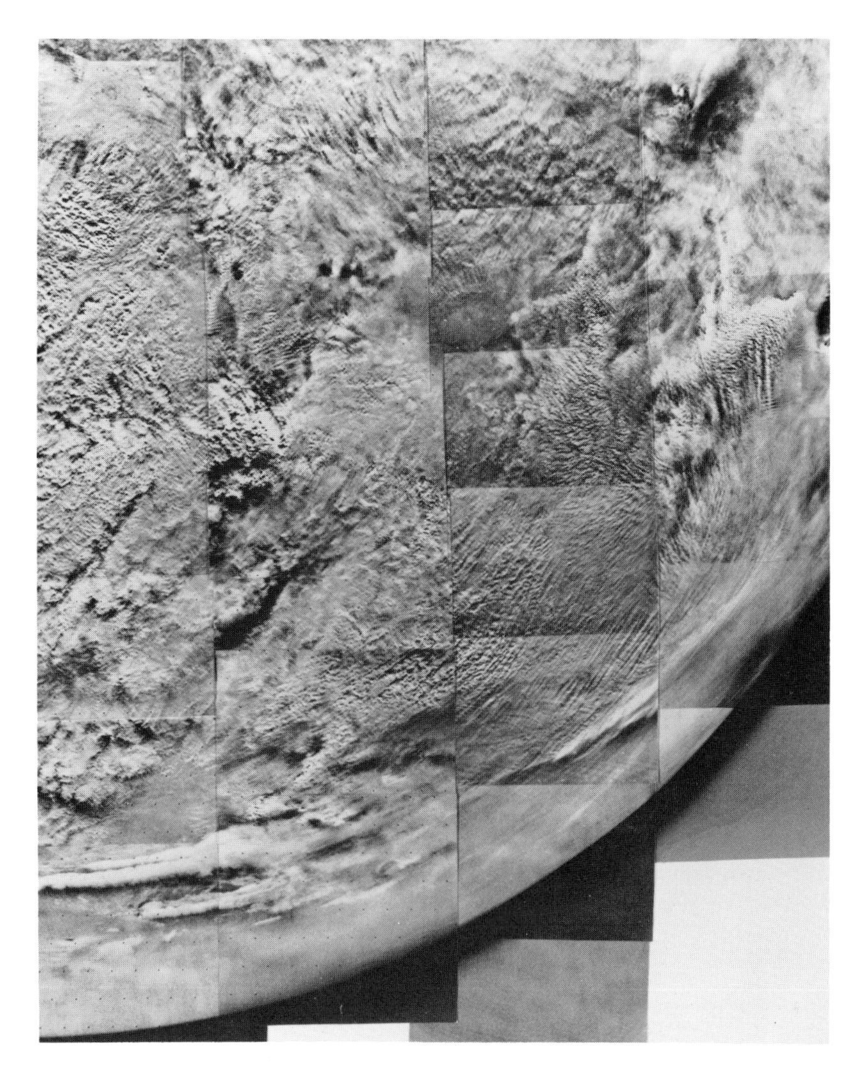

FIGURE 3.3: In this mosaic of images taken by the camera of *Viking 2* Orbiter on June 7, 1977, most of Mars' southern hemisphere is covered by a developing global dust storm which began a few days earlier. The great canyons are close to the limb, their floors obscured by mist or dust clouds. (NASA/JPL)

areas produces the seasonal albedo changes that have been seen on the planet since Mars was first observed by telescope and were formerly thought to be evidence of vegetation. Winds also contribute to features on the planet as well as the albedo markings. Streamlined erosional wind forms known as yardangs are larger than their terrestrial desert counterparts, and the winds sculpture enormous dune fields (figure 3.5) in equatorial regions and around the polar terrains. There may, also, have been eolian sculpturing of large rocky masses to produce blocky terrain with structures approximating geometrical shapes.

There is an astonishing geological diversity on Mars (figure 3.6), especially in the northern hemisphere which possesses vast areas of plains, an extensive region of chasms, and major volcanoes, bigger than any on Earth. By contrast, the southern hemisphere is dominated by heavily cratered terrain somewhat similar to the highland areas of the Moon, and by several large impact basins.

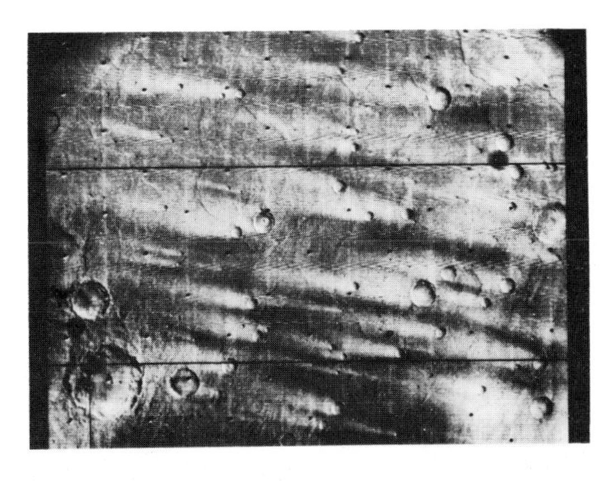

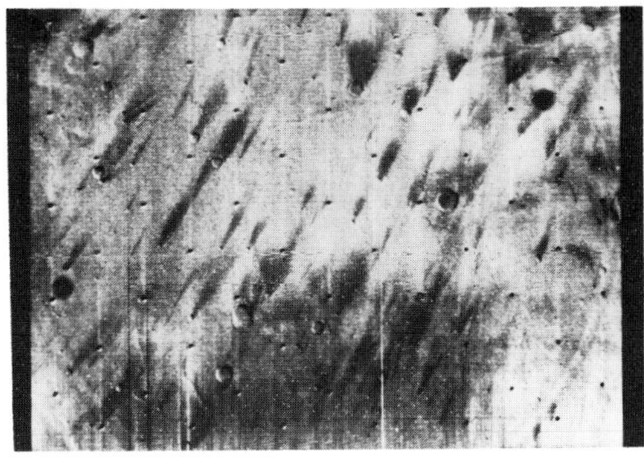

FIGURE 3.4: Two images from *Mariner 9* showing light streaks (top) and dark streaks (bottom) on the Martian surface. These streaks are believed to be formed by winds removing or depositing dust. (NASA/JPL)

The heavily cratered region differs from similar areas on the Moon in that the larger craters have been eroded and lost much of their sharpness (figure 3.7). There are volcanic intercrater plains and many craters have their floors filled with sediments, dune fields or lava flows. The area, too, has many drainage channels which suggest several pluvial periods early in the history of Mars and before the heavy bombardment ended. As the bombardment faded the heavily cratered terrain was active volcanically, giving rise to lava tubes and channels, cinder cones, volcanic domes, lava flows with wrinkle ridges, and many small volcanoes.

About one quarter of the surface of Mars consists of the region known as the Tharsis uplift (figure 3.8). This roughly circular area has a diameter of about 5000 miles (8000 km) and rises to 3 miles (5 km) above the mean surface level of the planet. It is dominated by a system of monstrous shield volcanoes. The largest, Olympus Mons, is 310 miles (500 km) wide at its base and its peak is 16.8 miles (27 km) above the mean level of the planet, three times as high as Mount Everest. Three other big shield volcanoes are in the same Tharsis area, together with several smaller volcanoes (figure 3.9). About 3000 miles (5000 km) to the east of Olympus Mons there is another somewhat smaller group of volcanoes in a region known as Elysium, also in the northern hemisphere. It is believed that the Martian volcanoes are so large compared with their terrestrial counterparts because the smaller Mars cooled more rapidly than Earth and developed a thicker crust. Plate tectonics could not occur on Mars and upwelling lava

FIGURE 3.5: This mosaic of *Viking* images is part of the Mangala Vallis region at about 8 degrees south latitude and 150 degrees west longitude. The images were taken June 19 through 21, 1980, four years after the spacecraft arrived at Mars. The northern part of the mosaic (top) shows an area originally covered by lava flows and later covered with wind-blown dusty or sandy material. Wind erosion is now stripping away the dusty cover and pitting and scratching the more consolidated material below. At the edge of the dusty cover, near the center of the mosaic, is a crater partially filled by the dusty material.

would continue surfacing at the same place to produce much greater shields. On Earth, by contrast, the movement of crustal plates spread the vulcanism over larger areas and, as is exemplified in the Hawaiian Islands, produced a chain of volcanoes instead of a single large volcano. Moreover, the thinner crust of Earth allows terrestrial volcanoes to sink back into equilibrium and lose height more than their Martian counterparts. The Martian volcanoes may still provide heat energy for anaerobic bacteria living below the inhospitable surface of the planet.

The rise of the Tharsis bulge strained the surface of Mars and produced an extensive system of faults from which an impressive canyon system was derived (see figure 3.8). Appearing like a great rift in the planet, the Valles Marineris (figure 3.10a) consists of enormous chasms stretching some 2500 miles (4000 km) with a width of 120 miles (200 km) at some parts. Canyon floors are about 10,000 ft (3 km) below the surrounding

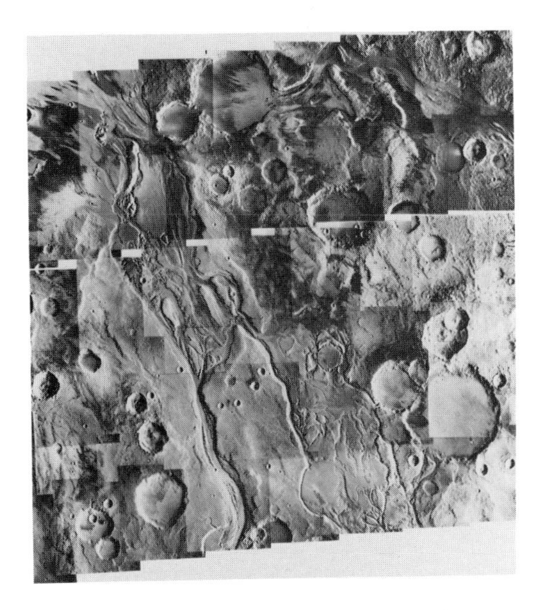
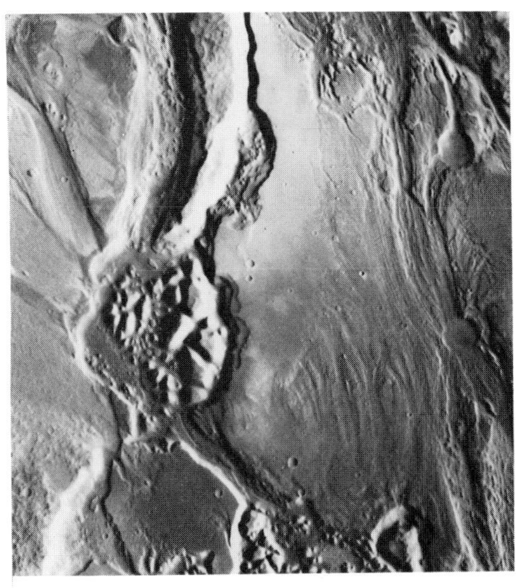

a, left
b, right

FIGURE 3.6: a) This mosaic shows another part of the Mangala Vallis region noted for its variety of geological processes. The central region contains vast channel systems that appear to have been carved by running water. Also there is evidence of many periods of deposition and erosion of surface materials by processes that include wind, mass wasting perhaps caused by water percolating beneath the surface, and fluvial action (the flow of water across the surface). b) An enlargement of part of the region showing details of the large flow channel. It shows mass wasting (the jumbled blocks), erosion by winds, and layering that suggests many episodes of deposition and erosion. (NASA/JPL)

terrain (figure 3.10b). It has been suggested that this rift was the beginning of plate tectonics early in the history of the planet, and that rapid cooling of Mars thickened the crust to a point where further plate movement became impossible. The floors of these canyons may be deep enough for water to be available at times for life forms there, either close to the surface or in alluvial deposits.

Mars has many dried 'river' beds (figure 3.11). Some are as much as 1250 miles (2000 km) long. They all appear to be very old, predating the volcanic region of Tharsis, and mainly located in the heavily cratered terrain (figure 3.12). Additionally, there are outflow channels, which seem to be much younger (figure 3.13). They are also much larger than the dry valleys. The outflow channels often start in areas of chaotic terrain where it appears that water melted from rock-ice mixes, possibly by geothermal heating. There are scour patterns and streamlining on the floors of the channels and many complex interconnections between channels. Often the channels breach craters and mountain ranges and produce large teardrop-shaped islands. Also, there is evidence of outflow channels from craters (figure 3.14) which would require a flow of water greater than could have been stored in the crater alone. These features seem to be the result of catastrophic floods which swept hundreds of miles across the planet's surface. The water from these floods may have gathered into shallow oceans on the northern plains where most of the outflow channels terminate. The water is now probably locked in thick layers of permafrost beneath the sediments and dust of these plains.

The *Viking* results are a basis for planning future biological exploration of Mars. What led to the selection of the *Viking* payload? It came after the National Academy of Science summer study in 1964. In the late 1960s and 1970s there was a period of

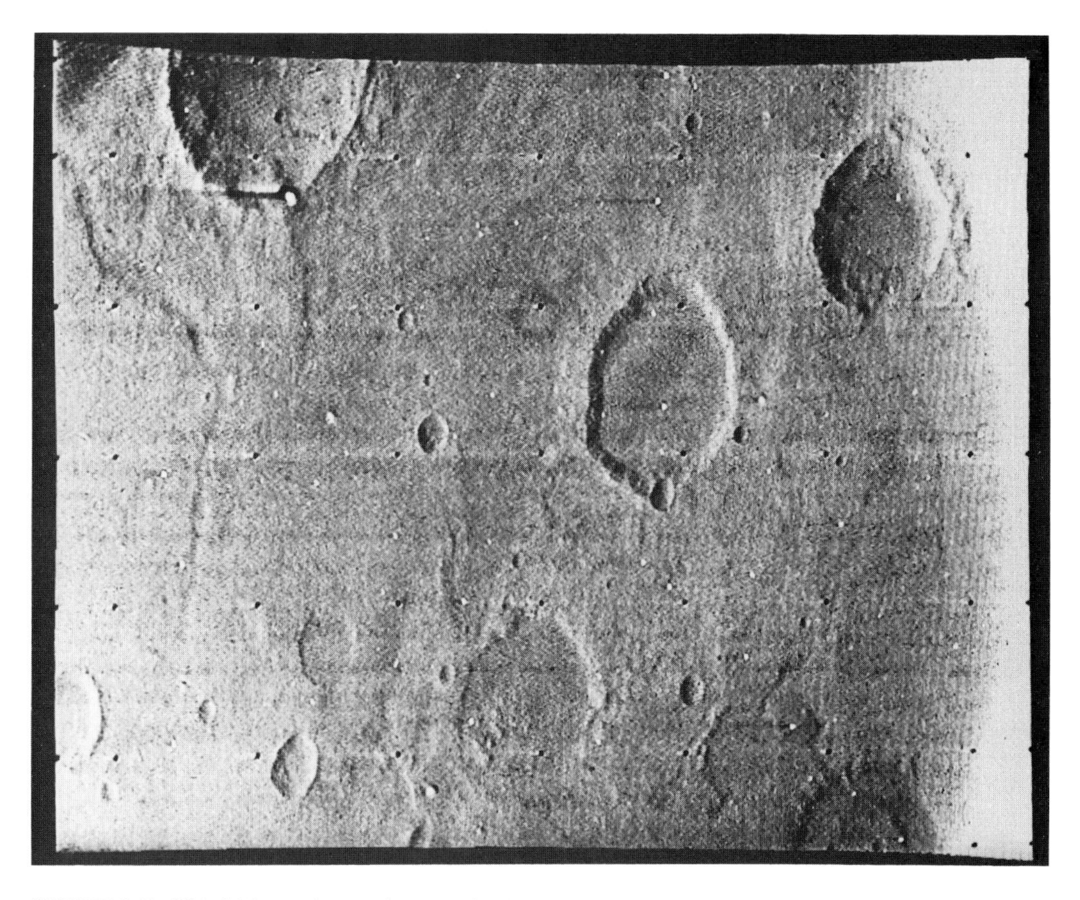

FIGURE 3.7: This high-resolution photograph taken by *Mariner 6* shows how large craters on Mars have lost their pristine sharpness. These craters appear quite different from their counterparts in the heavily cratered areas of Earth's Moon. (NASA/JPL)

intensive laboratory work to seek instruments that would search for life on Mars. Many ideas were put forward on how to go about looking for Martian life.

The search was based on two assumptions. The general attributes of terrestrial life, and what Martian life might be like. One extreme was that we could make minimal assumptions about what Martian life might be like, and assume that Martians would be like terrestrials, i.e., carbon-based chemistry. If the Martian life forms were carbon-based and similar to terrestrial life forms we could do enzyme analysis, look for specific biochemical markers, and also look for metabolism. We could look for organic compounds, for macromolecules, for optical activity, gas disequilibrium and metabolism.

Imaging by the *Viking* cameras could look for life on the surface. Pictures could be conclusive, but on an alien world they could be misleading, especially on a small scale. On Earth there are inanimate objects that look very much like living things, and living things that look very much like inanimate objects. For example, microspheres look very much like dividing cells but are quite inanimate. A South African plant, *Mesembryanthemum pseudotruncatellum,* resembles a small pebble except when flowering.

A seismometer on *Viking* recorded wind only, there were no vibrations from moving objects such as an animal might make. The cameras on the Lander showed no evidence of living materials at the landing sites, but this conclusion is now being challenged. Experiments on *Viking* incubated Martian soil to look for evidence of biological activity.

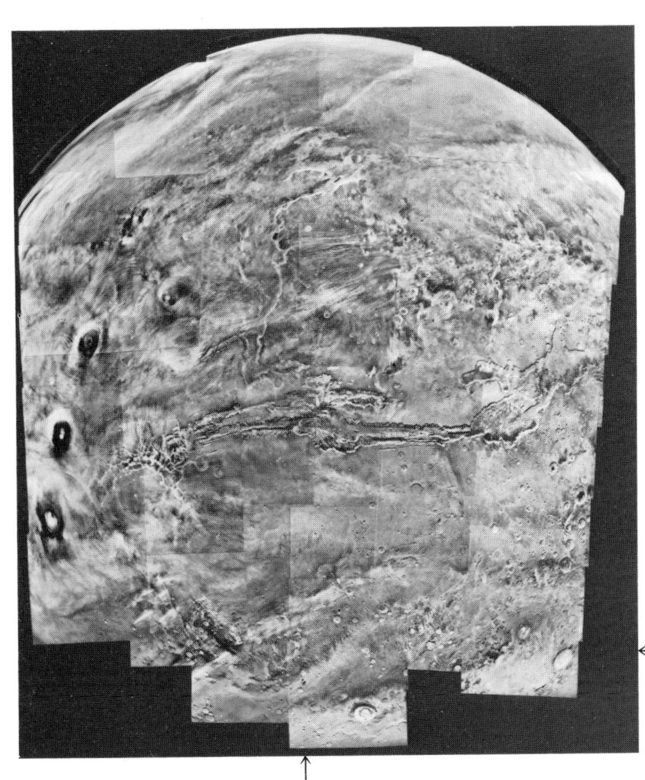

a) right
b) top left
c) bottom left

FIGURE 3.8: a) This view of Mars contains 102 images mosaicked together. It shows several prominent Martian features and at least two rare weather phenomena. Valles Marineris stretches across the center of the mosaic. The three volcanoes, Arsia Mons (bottom), Pavonis Mons, and Ascraeus Mons are at the left of the view. Curving north and east from Arsia Mons is a sharp line believed to be a weather front or a atmospheric shock wave (see also figure 2.5). In the southernmost frame there are three very small white clouds (identified by arrows at the margins of the picture), and slightly below (i.e., to the south) are their very distinct black shadows on the Martian surface. The shadows are unusually sharp. The largest of the three clouds is 20 miles (32 km) long and is 91,000 feet (28 km) above the surface. (NASA/JPL) b) Computer enlargement of the three unusual Martian clouds. c) An enlargement of the southernmost frame.

A gas chromatograph mass spectrometer did not detect any organic compounds. A gas exchange experiment showed an enormous and unexpected amount of oxygen release. A labeled release experiment showed no traces of metabolism by living things.

In the gas exchange experiments with its "chicken soup" incubation the quick release of oxygen and slow evolution of carbon dioxide possibly resulted from an oxidant in the soil which is not heat sensitive. The release of oxygen during the experiment was not affected by heat sterilization of the sample, nor by storage of the sample up to five months in the spacecraft before testing it. Thus the material responsible was heat stable and stable when held in storage. The observed reaction does not appear to be biological but a rapid reaction between water and a stable oxidant in the Martian soil. This oxidant is likely to be a peroxide, but not hydrogen peroxide. Also the experimenters found that there was less oxygen released from samples of soil which had an initial high water content.

In the labeled release experiment there was a rapid release of carbon dioxide which then plateaud. Heat sterilization of the sample before the test prevented any reaction, so it was shown to be a heat sensitive reaction which continues slowly for a long time. Storing the sample also resulted in no reaction when the sample was later tested. The

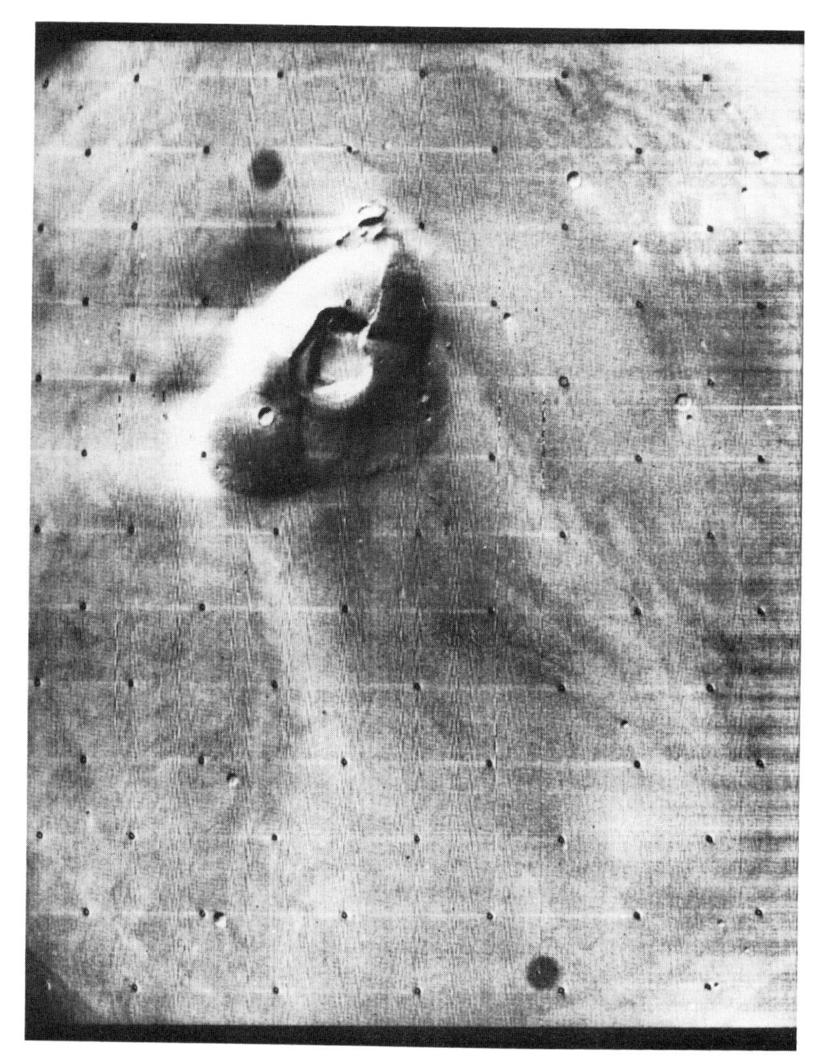

FIGURE 3.9: A relatively small Martian volcano, Tharsis Tholus, on the northeastern slope of the Tharsis bulge. It has not developed a huge shield like the big Tharsis volcanoes, but has a complex, although smaller, summit caldera. (NASA/JPL)

results could be explained by the presence of a biological system, and the experimenter says the experiments indicate the presence of life in the Martian soil. Other scientists disagree, pointing out that the rate of release would need a high concentration of living things of which no traces were found.

The pyrolytic release experiment resulted in the synthesis of organic material. The experiment was run nine times at two different sites on Mars with light on and off and some with water vapor present. Almost always the experiment produced a second peak of organic synthesis, but it was not photosynthesis because it occurred in light or in darkness. Water vapor is needed to help photosynthesis on Earth, but addition of water vapor to the Martian samples before the tests actually abolished the reaction. At 90 degrees heating, there was no effect on the second peak. Heating to higher temperatures did, however, reduce the second peak. It is concluded that we were not seeing a biological fixation of carbon dioxide which the experiment was designed to detect. Tests

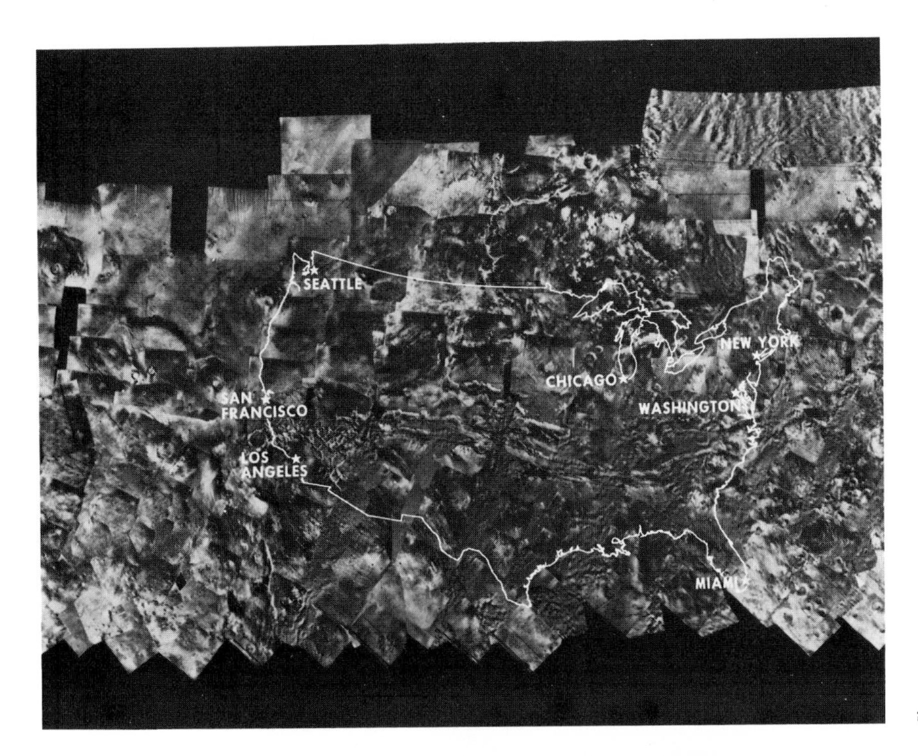

a

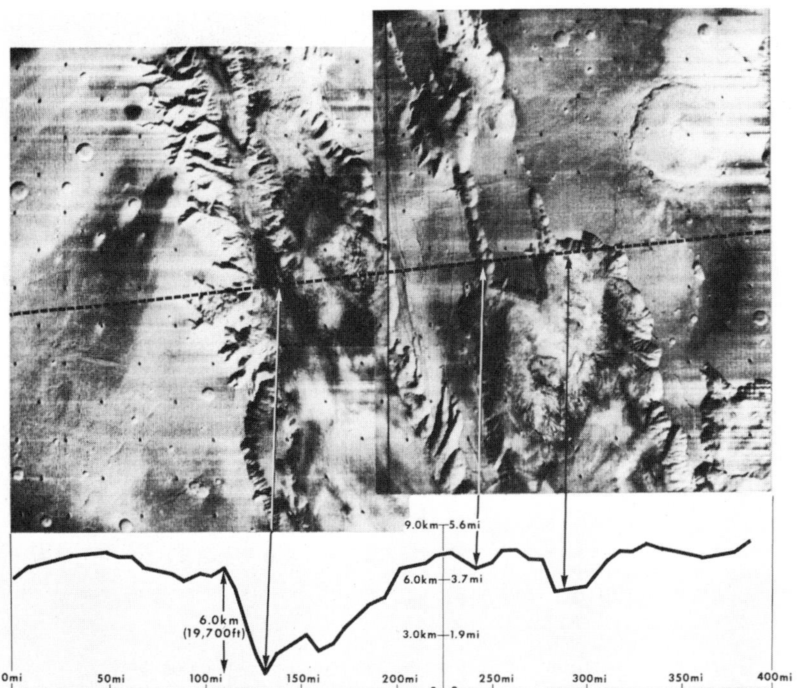

b

FIGURE 3.10: a) A panoramic view of one-third of the equatorial region of Mars is shown in this mosaic of images taken by *Mariner 9.* It covers longitudes from 10° to 140°, and latitudes from 30° South to 30° North. The region includes the great Martian canyons the size of which can be judged from the outline map of the U.S. superimposed on the mosaic. (NASA/JPL)
b) Two images of the Tithonius Lacus region of Martian canyons with a profile plot derived from pressure measurements. These vast chasms and branching canyons represent a landform evolution unique to Mars. (NASA/JPL)

FIGURE 3.11: This small channel system is located at 12 degrees north latitude and 142 degrees west longitude. The channel is 1.5 miles (2.5 km) wide and shows flow features along its length. The tributaries appear to have been produced by water seeping from beneath the surface rather than by surface runoff. Interruptions in the ejecta blanket from the large crater at the top of the image suggest that the channel was cut after the formation of the crater. (NASA/JPL)

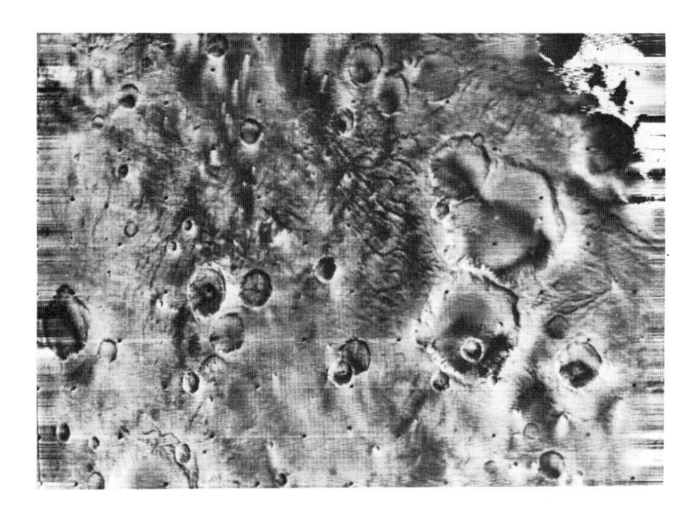

FIGURE 3.12: Heavily cratered ancient terrain of Mars has been modified by many small channels which are currently believed to have been cut by running water early in the history of the planet, possibly as a result of rainfall from a much denser atmosphere when there were also standing bodies of water on the planet. (NASA/JPL)

here on Earth suggest that the Martian soil contains a catalyst that is able to synthesize small amounts of organic materials.

In summary, the *Viking* data narrowed the possibilities of there being life today on the surface of Mars, certainly the consensus seems to be that there is no clear evidence for microbial life in the surface soils at the two landing sites.

There are terrestrial analogs of the present Martian environment. For example, regions of Antarctica (figure 3.15) look like the western deserts of the United States,

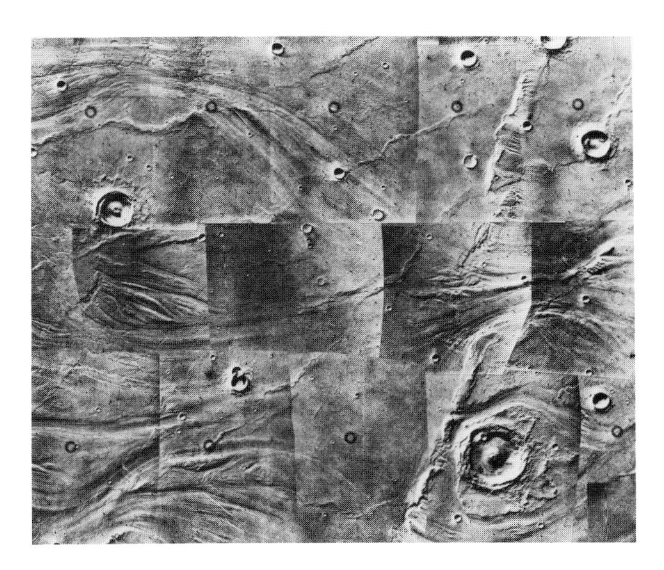

FIGURE 3.13: A different type of channel was formed later in the history of the planet when catastrophic floods poured toward low-lying areas. In places the floodwaters appear to have ponded behind ridges before cutting through them to erode significant gaps. Wrinkle ridges of the type seen on lunar lava plains have been modified by the rushing waters. The scale of the floods can be appreciated from the area of this mosaic; namely, 155 by 120 miles (250 by 200 km). (NASA/JPL)

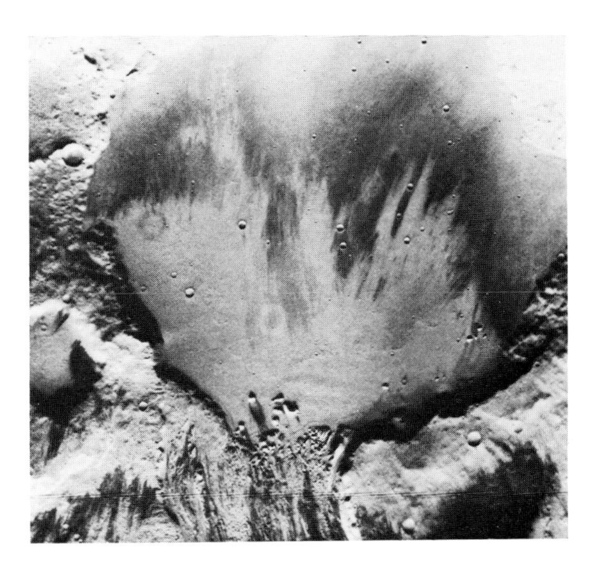

FIGURE 3.14: This crater in the Mangala Vallis region is 17 miles (27.5 km) in diameter. The mountainous crater wall at the bottom of the picture is broken by an outflow channel. To cut a channel of this size and depth would require a greater amount of water than could have been contained within the crater. Where all the water came from is unclear. Possibly it came via the major channel system shown in figure 3.6(a) on which this large crater appears at the right edge just above the center.

but they are bitterly cold deserts, although not as cold as the surface of Mars. Conditions on Mars favor chemical weathering: refractory minerals, water ice, and sulphates. Martian soil minerology may be very complex. Martian dry valleys are probably very much the same as Antarctica dry valleys, with a surface consisting of an eolian zone beneath which is a salt zone. Deeper still there is probably a chemically active zone,

FIGURE 3.15: A close analog of Mars is found in the dry valleys of Antarctica. Here, at Lake Hoare, scientists are studying life forms in the rocks and beneath the ice caps of lakes. (NASA-Ames)

then a seasonally frozen zone and a permanently frozen zone. All the zones are probably within a layer some 20 cm thick. Zeolites could be holding water in these Martian soils.

Prospect Mesa in Antarctica's Wright Valley has soils that are 15 million years old. The permanently frozen zone in this mesa starts at about 14 inches (35 cm) depth. Soils near the surface are enriched with salts. A 1-meter deep pit dug into the permafrost revealed a light colored salt layer near the surface. Collected cores showed that sulfur was concentrated at the surface, similar to what *Viking* found on Mars. Deeper there was chlorine, also similar to the chlorine found on the *Viking* surface samples.

Mars exhibits playa-like surfaces similar to terrestrial polar deserts. At the *Viking 2* site there were evaporate deposits, and there were incrustations on the soil surface at the *Viking 1* site, similar to incrustations in Antarctic dry valleys. These valleys, like the surface of Mars, do not receive any rainfall, but they have frost deposits similar to winter frost deposited at the *Viking* lander sites on Mars.

The Ross Desert valleys look remarkably like the *Viking* landing sites where we did not find microorganisms, but in Antarctica while there is no microbial life on the surface, microorganisms live in the rocks and at the bottoms of frozen lakes; the rocks and lakes are alive. A desert in Israel also has microorganisms in the rocks; actually the same as those in the rocks of Antarctica.

Cryptoendolithic lichens weaken the sandstone rocks leading to extrafoliation which reveals the active microbes below the surface of the rocks. This weathering produces a distinctive mosaic on the rocks which is indicative of a community of organisms below the surface.

In the Antarctic valleys ambient temperature only rarely exceeds 0° C, and it is nearly always below that level. How can the organisms survive in such a low temperature? What is the climate inside the rock? The microorganisms colonize the sunny side of a rock, while the other, shaded, side does not show microbial weathering. In addition, melting snow or frost on the rocks provides water which is maintained within them for a long time as liquid surface films around sand grains, even at a temperature of −8°C.

It is important to note that at the *Viking 2* landing site in Utopia, a thin layer of water ice coated rocks and soil during winter. The layer was estimated to be about .001 inches thick, but would be capable of providing water for any microbes within the Martian rocks.

In Antarctica, the ambient temperature is above $-5°$ C for only 30 days each year. Sunlight provides the source of heat which internally maintains the rocks at $-7°$ C. When scientists wrapped up the rocks in layers of window screens to find the effects of reduced light they discovered that when only one-quarter of incident sunlight reached the surface of the rocks, photosynthesis by the microorganisms leveled off. But metabolism continued slowly even at $-8°$ C. When the temperature reached $-5°$ C metabolism started to increase until it reached a peak and, as the temperature continued to rise, metabolism became inhibited when it exceeded 15° C.

The controlling factors for life within the rocks are water, light, temperature, nitrates, and carbon dioxide. Carbon dioxide is produced at temperatures between -5 and $-1°$ C, so these organisms live at the absolute margin of existence within a zone of life which is 2 to 10 mm inside the rock.

Living on the fringe of life these creatures recycle carbon in a period ranging from 576 to 2350 years, compared with 0.7 to 27 years for microbes in other terrestrial habitats. If conditions changed only a degree they would not survive. This is why these rock microbes are most likely to be similar to the last vestige of life on a dying Mars. In fact there are both living and dead rocks in Antarctica, and evidence of the dead communities remain in some of the rocks. The organisms change the rocks sufficiently for scientists to identify past microbial activity. The same may be true on Mars.

Another terrestrial analog of where evidence of Martian life might be found is the biology of anhydrobiosis, creatures that can survive for long periods without water and in a completely dehydrated form. Under some circumstances plants and animals can lose body water, dry up, and later be brought back to life with water. One type is where this dehydration is a normal part of the life cycle. An example is the brine shrimp. The female lays cysts, which remain dry for years. When hydrated they hatch out as larva. Spores are another example of a normal life cycle which includes a dehydrated phase of life.

Another type is where the dehydration takes place at an adult stage of the organism and it can occur repeatedly. An example is the tartagrade. This creature can contract into a barrel shape and can even exist at low temperature in a vacuum. Some nematodes also fall into this class.

Clearly all metabolism stops when the creatures are dehydrated. It is unlikely that in their dehydrated form these creatures could have any metabolism because there is no water chemistry and no observed enzymatic activity. Yet life has been defined as the characteristic of having an organized structure, not necessarily having metabolism. However, without metabolism the organism cannot make any repairs to its DNA, so the dehydrated organism may not be able to survive indefinitely in its dehydrated state. On Earth we know that Manchuria lotus seed has survived for 1000 years, tartagrade and rotifer nematodes for 120 years, and Arctic lupine for some 10,000 years.

Survival is also maximized if the creature is exposed to high humidity before dehydration. But drying out has to take place slowly so that the creature can change morphologically. As the creatures dry out, membranes crystalize to gel, and lipids and proteins group together in an entirely different pattern from the normal hydrated form of the organism. Rehydration redistributes the lipids and proteins, and these then become arranged in the same pattern they occupied before dehydration, the membranes return to their normal form and metabolism is resumed.

Identifying such creatures on Mars, if any remained as the planet freeze dried, may be an extremely difficult, though not impossible, task. It most likely cannot be attempted until we have a permanent manned base on the planet.

Could life exist on Mars today? In the late 1960s Dr. Richard S. Young of NASA simulated Martian conditions of pressure, temperature, and radiation and introduced terrestrial organisms into this environment together with a small quantity of water. He found that one bacterium, a rod-shaped pseudonomas, grew well in the simulated Martian environment. After several generations it even survived the chilling night-time temperature of $-175°$ C. Another aerobachter bacterium, aerogenes, ignored long periods of freezing and still increased its population. Terrestrial life forms have a fantastic power to adapt to alien conditions. Many can grow and reproduce over a range of temperatures from -18 to $104°$ C. Some survive the pressure of our deepest oceans and the rarefied atmosphere at the tops of the highest mountains. They can be active in distilled water and in saturated brine solutions.

Halobacteria can grow on a salt crystal and compete with the crystal for water. Such a bacterium uses as much energy to do this as it would have to use to extract water from the atmosphere of Mars.

In 1971, I interviewed Dr. Sanford M. Siegel at the University of Hawaii. He had showed that terrestrial life can thrive in many environments previously thought to be inimical to life. In soil samples from North Wales he had found a small living organism that thrives when placed in an ammonia-rich atmosphere; it has survived the oxygen contamination of Earth's atmosphere by burying itself deep in the soil. Working with seeds which he suggested are rest points in the life of an organism, he showed that seeds which germinate in a normal oxygen-rich atmosphere cannot do without oxygen as plants, but that seeds which germinate in an oxygen-free atmosphere cannot later tolerate an oxygen-rich atmosphere. Dr. Siegel also reported that he had maintained spores in liquid ammonia until they started to metabolize and produced nucleic acid and proteins as though they had adapted to using liquid ammonia instead of water in their life process.

He cited the case of a Navy reconnaissance flight discovering an unusual pond in Antarctica, literally a brine puddle whose edges were encrusted with chloride crystals which was named the San Juan Pond. The water was always colder than that at which normal life forms live, the midsummer temperature was less than $-3°$ C, but still the pond contained bacteria and molds. Dr. Seigel reproduced these conditions in his laboratory and was able within a few months to have colonies of algae, molds, and bacteria living in the laboratory freezers.

It appears that once prebiotic molecules are on a planet they evolve into complex building blocks of organic matter, triggered by energy. The next step of evolution into organic life is still a great mystery. The transition stage is not recorded in the fossils of Earth. We may discover it on Mars. And once life has evolved it appears to be extremely tenacious in continuing in spite of changes to the environment. A major question is whether or not life evolved on Mars, and whether or not it was able to change in ways that would allow it to adapt to the changing environment on the Red Planet. Only a thorough examination of Mars can provide answers to these questions.

In searching for evidence of earlier life that most likely developed soon after Mars accreted and possibly during the tail end of the heavy bombardment from space, unmanned or manned expeditions should go to valleys and mud flow craters. Soil samples should be obtained from deep below the Martian surface, from the side walls of canyons, from subsidence valleys, from old river beds, and from old sedimentary rocks, collection ponds, and valley systems. For example, a good place to search would be in the bottoms and side walls of the canyons of Valles Marineris.

For a site to be suitable it must satisfy several criteria. It must contain or have contained water. It should be ancient and should not be oxidized. It should also have undergone limited metamorphosis since the time when life might have been present, and it should be accessible to manned or unmanned rover vehicles.

There may be areas on Mars conducive to the existence of life today. Anaerobic bacteria may still thrive on volcanic heat below the surface. Soviet scientists are optimistic that they will obtain data to lead them to oases where Martian life still thrives. U.S. scientists, by contrast, think this search has a low probability of being successful. Moreover, the U.S. scientists claim that the search for extant life on Mars is not a valid reason for further exploration of the Red Planet. More important is the quest to understand how planets evolve, how life originates on planets, and how planets change to becoming inhospitable to life. We should be trying to find evidence of earlier life to throw light on the important question of how living things develop from prebiotic molecules and how they adapt to catastrophic changes to the environment.

Scientists have developed a comprehensive theory of chemical evolution, but this is based on observations of terrestrial life. For the theory to be accepted as universal we need to get evidence of whether or not life evolved on other planets. If we could prove life originated on another planet this would also have a big impact on the search for extraterrestrial intelligence. Finding evidence that life originated elsewhere even if not still there today is a worthwhile objective in itself.

The Moon, although relatively easily reached and explored from Earth, is not a suitable body on which to look for early life in the Solar System. The Moon is too small to have ever possessed liquid water on its surface or an appreciable atmosphere after it accreted and went through the period of intense bombardment from space. Neither is Mercury a suitable planet for life to have been able to get a foothold. It also is too small. Venus may have possessed oceans in the past, and life may have started, but if so it would have been quickly snuffed out as the planet went into a runaway greenhouse condition and lost all its water. Today the conditions are so inimical to life that we would find it completely impractical to search for evidence of extinct life there. As for the outer Solar System where planets and satellites have good supplies of the volatile elements so essential to life as we know it, all are too cold or too gaseous for life to be probable or a search to be practical. Except for speculation that life may exist in ice-covered water oceans of Jupiter's satellite Europa, the remaining planet is Mars.

Mars is a fossil planet and may have had life. We need to know if and where there was water on Mars. Where was its source and where did it go? Water is the important question because water is essential for life as we know it. In fact life depends on water chemistry even more than on carbon chemistry. Some scientists strongly believe that terrestrial life began in clays.

A major breakthrough in explaining the origin of life appeared to have been made in 1977 when scientists at NASA-Ames Research Center showed how basic types of life molecules—amino acids and nucleotides—could have been concentrated in primitive oceans of Earth. Clays were most probably widespread on the primordial Earth and the shores of its oceans. Dr. James Lawless and his team of scientists found that when low-concentration solutions of amino acids were mixed with commonplace metal clays the clays attracted amino acids out of solution. This was especially so of an abundant nickel-containing clay which preferentially attracted the 20 amino acids which form the proteins of living cells. Other clays were found to destroy non-protein-forming amino acids faster than protein-forming amino acids. Thus, a realistic mechanism for the concentration and selection of life-forming amino acids on a water-rich planet had been discovered.

Metal clays have similar effects on the building blocks of DNA, the long chain molecule carrying the genetic material of living cells. DNA building blocks are concentrated by zinc clays. Zinc plays a significant role in action of the enzyme, DNA polymerase, which performs the catalytic task of linking DNA building blocks, the nucleotides of living cells. The work seems to show how amino acids not suitable for life were selectively destroyed while life-related amino acids were linked together in the ancient oceans to produce the chains needed to make biological cells and to produce a genetic blueprint for living organisms to reproduce and multiply.

We are sure that there are clays on Mars. Scientists want to return to Mars to look for where water was and now is, and then find clay deposits and sediments and there look for evidence of past life and organics.

Only comparatively recently have scientists discovered that here on Earth life has been around for most of the planet's history. We did not know this when the *Viking* missions were planned. Also we know that, even today, most life on Earth consists of single cells. So the strategy of the search for life on Mars must be revised from that mounted on the *Viking*s. We should look not necessarily for the microbes themselves but for structures built by single cell organisms. But where do we look? Microbes form colonies in coastal waters of Earth. Similar structures from ancient microbial colonies might be found in areas where there were shallow waters on Mars.

Some 3 billion years ago or earlier, when there was water on Mars, we know there was life on Earth. Hence there is a high probability that there was also life on Mars.

If there are microbes living on Mars today they might be anaerobes which do not depend on oxygen. They would be expected to function at a low metabolic rate and be long lived, possibly quickly consuming any dead microbes. Much of the time they would be dormant as are many terrestrial life forms that live in hostile environments. They must also be resistant to extreme cold and to dryness. Since conditions have been so inimical to them, such microbes would not be expected to evolve but would remain virtually unchanged for many millions of years. Higher forms of life may never have evolved on Mars.

An important aspect of any future mission to Mars should be to look for life or evidences of life as well as test for it. It may be easier to see evidence of living things or their activities than test for them chemically as was the case with Viking. But care will have to be taken in rejecting "analogs" that are not biological.

The capacity of life to survive harsh conditions should not be underestimated, especially as it was most likely that the change to today's harsh conditions on Mars from the benign conditions of an earlier age was gradual, possibly taking many millions of years. On Earth we have examples in Antarctica of how life adapted to the slow change from subtropical to arctic over millions of years. Today life continues beneath the surface of rocks and beneath ice caps on deep lakes.

Fairly recently scientists have discovered that even the winter sea ice in the Antarctic, formerly thought to be devoid of life, is actually teeming with living things. A planktonic crustacean called krill spends the winter in holes within the ice which acts as a labyrinthine nursery for the developing creatures. They feed on algae which in turn use sunlight penetrating the ice as their energy source. In fact, the ice acts as a protective shield preventing predators from eating the krill while they are developing into adults. Also, many single called organisms in the ice labyrinth feed on the algae.

Mars had a climate similar to Earth at the time several billion years ago when life evolved here. Also Mars has older rocks than Earth which will permit a search for the origin of life from prebiotic molecules. Inorganic clays are believed to have preceded biolife on Earth. Indeed, it may be more productive to look for clays rather than for microbes on Mars. Life is a planetary phenomenon, and an important question is: if, when, and why did life end on Mars?

It is highly improbable that life is present on Mars today, but highly likely that the planet will hold a record of life. There is strong interest in the U.S. and the USSR by scientists who want a return to Mars to explore the planet in more detail. The question Is there life on Mars? has been partly answered by *Viking.* A more important question —how does life evolve?—can be answered only by examining another planet in addition to Earth. And Mars is the most suitable planetary candidate for answering this question. As one scientist stated at a recent press conference on the evolution of life on Mars: "It is better to cooperate with the Russians in search for life on Mars than compete with them in ending life on Earth." Mars has as much land area as our Earth because much of the Earth is covered by oceans. The big question is where to start the search. In looking for both extinct or extant life on Mars we would first hunt for water, and the most likely places to do this are the polar regions, slushy areas, subsurface waters, aquifers, and springs. Those are the places where the evidence of Martian life is most likely to be found.

In the development of life on Mars we would expect to have had water acting on clays to produce cyanobacteria and other bacteria living in sediment rich layers. Life in a sediment responds to sunlight. Soon after the conditions developed on Earth for the origin of life on Earth, the cyanobacteria appeared. They lived off sedimentary surfaces in shallow water and adapted to ultraviolet perhaps through a DNA repair mechanism. Perhaps the sediments provided a protective cover against solar ultraviolet or the bacteria developed a mucal protective sheet. Certainly by 3.5 billion years ago these creatures were active on Earth and were probably active on Mars also. By that time all the major groups of microbial terrestrial life forms had evolved, and since then not much has changed as far as these creatures are concerned. On our planet evolution progressed very rapidly over a period of less than 400 million years. For example, all the major plans of the phyla had evolved by 3.5 billion years ago. The phyla of marine invertebrates all originated within 100 million years or so and in the following billions of years no new phyla evolved although some became extinct. Nothing new developed, there was only a fine-tuning of the species.

On early Mars and Earth there would be a very unstable environment; sedimentary rocks being deposited, lava flowing, and meteors bombarding the surface. On Earth we find inorganic spheres originating from impacts in both deep water and shallow. These are silicate droplets quenched in water in which iridium levels are high, so we know they came from space. The impacts created tsunamis that rushed as great floods over what was then a fairly featureless Earth. But the planet had life thriving and adapting in inimical conditions of enormous lava flows, globe circling clouds, floods, and impacts —life adapting to an extremely hostile environment. Yet life adjusted to thrive on Earth despite all these difficulties. Martian history also was chaotic and life should have been able to adapt similarly. The question is how wet and how warm for how long were conditions on Mars? What we will be forced to sample on Mars in a search for life there will be debris, unless we can dig deeply below the surface.

Exobiologists must develop a strategy for all kinds of situations which will be set by mission planners. They must define the many options for exobiological science, but unfortunately exobiologists are most unlikely to be able to direct the mission. Nevertheless, they must still make their unique requirements understood when future missions to Mars are being planned.

A good site to explore Mars for life is a lake deposit. The search will probably have to be for stromatolite debris. There are what appear to be paleolake sediments on Mars; Valles Marineris, by way of example, has layered deposits which may be from a lake environment. These lakes on early Mars were most probably originally ice free but later became ice covered. The lakes might have been free of ice for 1 billion years, then ice covered for a further 2 billion years. The final Martian lakes were undoubtedly ice covered. If this is a correct interpretation of Martian history, the existence of such lakes would extend the time that life could exist on Mars. Sediments from these lakes may provide evidence of this early Martian life and will be good locations for searches.

We have a good analog of Mars on Earth. Antarctica is a cold desert like Mars is today and accordingly may be an ideal place to set up a Mars simulation station in preparation for a base on Mars, making the station "live off the land" by growing its own food and having its crew sealed in a self-sustaining environment, going outside only in Mars-type spacesuits. *Landsat* images of the dry valley region of Antarctica look very much like Mars as seen from a Mars-orbiting spacecraft. Also, the surface of the dry valleys looks very much like the surface of Mars at the *Viking* lander sites on Chryse and Utopia.

These dry valleys have perennially ice-covered lakes (figure 3.16). Scientists established a research base on the edge of one of these frozen lakes and found a 15 to 20 foot ice cap over 100 feet deep water. They found that 90 percent of the incident sunlight is lost, but 10 percent penetrates through the ice cover to the water below. There are no prehistoric fish, no insects, no currents to disturb the sediments, but cyanobacteria and diatoms; a consortium of organisms in a tissue type of matrix on the lake bottom. Photoautotrophy seems to drive a bio system beneath the ice, and microorganisms form mats on the floor of the lake. These microorganisms use pigments to trap the feeble light for energy.

Sediments from wind-blown surface material get down through cracks in the ice in episodes. The mean annual temperature is now −20° C, but it is becoming warmer, and the thickness of the ice has been decreasing over the last ten years. The ice cover acts as a lid to concentrate biologically important gases beneath it. The ice also acts as a thermal buffer. It thus provides a temperature and a gaseous environment suitable for life. The valleys have been periodically inundated so there is evidence of ancient microbial mats also.

Terrestrial playa lakes, the last remnants of large water basins, are also analogs for early Martian conditions, and their biology is another model for early Martian life. A large area of Central Australia has many such lakes, e.g. Lake Aire. These lakes are of two kinds, permanent and ephemeral. The latter have intermittent inflows, high evaporation, and catastrophic runoffs. They show evidence of minerals, microbial mats, and

FIGURE 3.16: In searching for life in the frozen lakes of Antarctica's dry valleys, scientists melted holes in the ice cover and deployed equipment to discover that microbial life was living in the waters of lakes such as Lake Hoare. Microbial life might also have adapted to live in frozen lakes on Mars as the Martian climate cooled. Evidence of such Martian life will be sought in deposits which are thought to be the sediments of ancient Martian lakes. (NASA-Ames)

intersedimentary salt growth. Sometimes they are almost dry and are then great salt flats with small high-salinity ponds in which some biological communities still function in salinities up to 400. Cyanobacterial mats operate in conditions varying from dessication to flooding. So they grow and then dessicate and repeat the cycle. Water uptake is very rapid followed by rising photosynthesic activity. The bacteria surround themselves with a protective sheath, whose pigments protect them from ultraviolet radiation. The sheath also acts as a dessicant buffer.

In any search for evidence of early life on Mars, similar lake beds should be sought, and one way to identify them may be by geological structures showing ground water resurgence features. In picking experiments to search for biology past and present on Mars, the following criteria are essential. First the sample must be checked to find out if it contains detectable amounts of organics or of reduced carbon. Developments from the Viking gas chromatograph mass spectrometer instrument can be used for this purpose. Next, the sample should contain water in a form accessible to living things. Access to liquid water is absolutely necessary for terrestrial life forms. Water vapor or ice are not suitable. Commercially available instruments can detect the amounts and phases of water in a sample. The next question is whether or not a sample contains water soluble electrolytes which are universally present in terrestrial life forms. These electrolytes play important roles in the operations of enzymes and other physiological functions of living things on our planet which make extensive use of ions of sodium, potassium, magnesium, and calcium, for example. Such electrolytes can be detected by placing regolith samples in water and checking on the resultant changes in electrical conductivity. Another important criterion is whether or not the gases in the sample of soil differ from those in the atmosphere, since biological activity always alters the composition of the gases surrounding the living organism. An instrument developed from the *Viking*'s gas chromatograph mass spectrometer can provide this measurement.

Places on Mars where it would be most advantageous to gather samples in the search for extant or extinct life include sediments or layered material such as alluvial flows, interfaces between ice and the regolith at the margin of the permanent polar cap or where the regolith lightly covers known areas of permafrost, and in cores derived from depths exceeding 1 foot (30 cm) beneath the surface. Unfortunately we do not know at present how deep the oxidants extend and how far solar ultraviolet radiation penetrates beneath the surface. While solar ultraviolet is unlikely to penetrate more than a few millimeters, its effects can be much deeper in destroying organics if the surface is turned over in depth, such as by wind erosion and dust deposition. So the depth from which we must take samples depends upon how deeply the surface material has been and continues to be mixed by weather conditions, dust storms, wind, and impacts, and, in the case of the oxidant, how much it can diffuse into the regolith. Also, extant life if it exists on Mars today would have to protect itself from high energy cosmic rays and from solar protons. This requires several feet of regolith as a shield from such damaging particles.

The most promising locations at which a search for life on Mars should be made appear to be those which might contain organic rich materials. These are sinks of materials which have been produced through chemical weathering of igneous rocks, basically the processes of oxidation, hydration, carbonation, and of salts entering into solutions. Many chemical weathering processes are generally irreversible processes, such as the conversion of water to the OH radical in minerals. Also, thermodynamics and kinetics are non uniform, such as differential leaching by water from minerals. Sinks are produced through adsorption and ion exchange. Examples are layered structured silicates ("clay" minerals) such as zeolites.

On Mars aluminosilicate adsorbents are most likely traps for organic compounds. An expedition to Mars should look for clay minerals. Sites with rusty materials would be unfavorable sites for organics because iron oxides are antagonistic to carbon compounds. Thus, soil units containing silicates but little rust are preferable sites for the search for biological materials.

The origin of life is a problem of chemistry and of environment and of changes to it. The known matter in the universe consists mainly of hydrogen and helium with relatively small amounts of other elements. Most abundant of these other elements are carbon, nitrogen, and oxygen, and the chemistry of the universe is mainly organic.

There is a strange irony in that the terrestrial planets appear to be the most likely places to have life based on solar heating, but they are the worst places to find organics. In the continental crust of Earth the abundance of nonorganic materials approximates that in the universe, but volatile elements used by living things are 100,000 times less than in the universe generally. It is ironic that most volatile elements are concentrated in the cold giant outer planets and their satellites, while the small inner planets are the worst places for organic chemistry.

An important and as yet unanswered question is how life began on Earth, a planet that is depleted in volatiles, that is, in carbon, nitrogen, and oxygen. Mars may hold the key to answering this question.

As discussed earlier in this chapter, when evidence of Martian life was sought with the biology instruments of the Viking landers researchers found that the soil synthesizes, the soil expires, but the soil contains no organic molecules. Most closely these conditions are modeled by an iron-rich clay produced by chemical weathering of igneous rocks. Clay produced by chemical weathering of igneous rocks is silicon surrounded by oxygen atoms. It is arranged as a two-dimensional tetrahedron sheet interspersed with octohedron sheets; it has a two-layered structure.

Over half of sedimentary rocks on Earth are clays. If water is liquid on Mars, clays would be expected to be abundant on Mars also, and as discussed earlier in this chapter the activities of clays were shown as a possible way that amino acids might have been selectively assembled to originate life on Earth. Iron-rich Martian clays may be responsible for the observed chemistry of the Martian soil. Clays can replicate themselves in a kind of template system. Also, they can not only replicate but also mutate, thus permitting evolution. In this respect clays may be where life originated rather than in a primordial "soup" as was suggested by earlier researchers.

Water produces replicating clay crystals. Ferrous iron is the major reductive material. Life needs to reduce carbon dioxide and requires light to do this. Ferrous iron and carbon dioxide and light lead to formic acid and to other molecules essential for the life processes. In the *Viking* experiments carbon dioxide was reduced through the action of light which suggests the presence in the Martian soil of a replicating clay possessing a memory system. Sunlight was a major factor in the development of terrestrial life and it was utilized advantageously by Earth organisms in their early evolution. Iron allows fixation of carbon dioxide with light, and it could be that iron-rich clays with the best survival became our life form. Was this also true on Mars?

We know that the atmosphere of Mars is carbon dioxide and nitrogen. The carbon dioxide needs to be reduced to acids, bases, sugars, and the like for its use by living things. Life is essentially based on water chemistry. Twenty-nine chemicals are involved as the fundamental molecules of life and it is the hydrolysis of these substances that produce the variety of life.

Liquid water seems to be essential if our form of life is to evolve on a planet. We believe that Mars once possessed large amounts of water in its liquid form. When liquid water disappeared from Mars by freezing or otherwise, it would seem invitable that life would also end on that planet. Today the iron-rich clays of Mars may be the current evidence of life that once thrived on the planet, life that evolved when Mars was water-rich early in its history. If life did exist on Mars exobiologists would like to know if for a time it followed similar evolutionary pathways as on Earth.

We can access the geological record on Earth back for 3.8 billion years, and we find evidence of early life being on Earth about 3.5 billion years ago. When searching in the field for evidence of early life on Earth it is essential to look for the proper rock types. Chert, an impermeable silicon rock, preserves microbes and bio-material. Shales or limestones do not. It is also essential to be sure that biological material found in ancient rocks is a true fossil and was not introduced at some later time.

Several categories of evidence for earliest terrestrial life have been discovered and identified. First, as mentioned earlier, precambrian terrestrial microfossils have been found beautifully preserved. While some limestone contains black nodules of chert with preserved specimens, there are no fossils in the limestone itself. In the 2 billion years old Gunflint cherts of Lake Superior there are abundant and well preserved microfossils in which there is evidence of cell division and reproduction. There are also bizarre forms of microfossils, for which there is no modern analogy. In 3.5 billion years old chert from Western Australia there are irregular layers with tubular structures and filaments and some partitioned tubular structures in which one filament is wrapped by another filament. And in cherts from Swaziland, South Africa, there are coccoids, and microfossils with threadlike and tubular filaments. Similar microfossils might be present

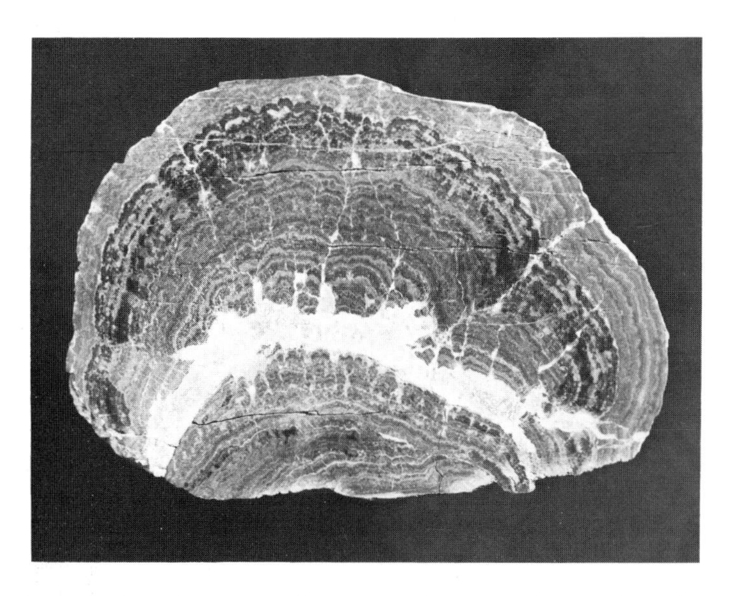

FIGURE 3.17: Cross section of a stromatolite; stromatolites are columnar layered sedimentary fossils formed by microbial life. They have been around on Earth for at least 3.5 billion years and were probably also on Mars during its wet period about the same time. Searching for such evidence of early Martian life will be an important task for the forthcoming manned missions to the Red Planet. (NASA-Ames)

in Martian rocks if we can decide where to search. This type of search must, however, await a permanent manned base on the planet because it requires processing of much material.

Another category is that of stromatolites which are organosedimentary structures produced by microrganisms—mainly cyanobacteria—which trapped, bound, or precipitated sedimentary materials (figure 3.17). They are easily recognizable as structures produced by microbes. Some can be dated as far back as 3.5 billion years ago, and some are still being made today in Australia and in the Bahamas. They are often built by microbes in a marine setting, but also they occur in lakes such as Lake Clifton in Western Australia, and Lake Tanganyika in the Great Rift Valley of East Africa.

Stromatolites became very prevalent about the time of the first animals on Earth. The sizes of stromatolites vary enormously. Some reefs are hundreds of miles in extent. Single stromatolites vary from hundreds of feet down to microscopic sizes. Rate of sedimentation and growth of stromatolites is in a delicate balance, for the rate of sedimentation must be less than the rate of growth. Also cementation is needed so that compaction does not later destroy the evidence of life having been present.

In Western Australia there is a 100-yard-wide zone of stromatolites along a coastline, but they are not very abundant. There are no stromatolites in salt lakes among dunes, they are concentrated in shoreline intertidal lagoons. The most extensive stromatolites consist of flat sheets of microorganisms, with a crinkly surface. The bacteria cement the sediments. In the intertidal zone they make rocklike nodules. In the subtidal zone they produce pillar structures, but these are rarer than the intertidal stromatolites.

Terrestrial stromatolites grow in river systems, too, where they produce dams. They also form in cold and hot springs where they are more common than in lakes and shorelines. Good examples are in Yellowstone National Park where hot springs are surrounded by microbial mats.

Stromatolites would be expected to have developed in Martian lakes and shorelines.

Hopefully a Martian rover could recognize Martian equivalents of these structures. This will be much easier than searching for rocks that might contain microfossils because paleontologists have to collect many samples to find microbial fossils, which would be difficult to do on Mars. However, on Earth stromatolite debris is found more commonly than stromatolites. So on Mars it may be desirable to look for such debris rather than search for the stromatolites themselves.

In searching for such features on Mars one has to be careful to avoid being confused or misled by analogs; for example, on Earth in arid regions, soils develop carbonate hummocks that are very similar in appearance to stromatolites and they are difficult to differentiate from true stromatolites. These inorganic knobs look very much like stromatolites, and the layered interior section is indistinguishable from stromatolites. A manned expedition to Mars may, however, be able to use the isotope composition of carbon and oxygen to differentiate between carbonates and stromatolites.

A 2-billion-year-old area of Western Australia looks very much like Mars, with river beds and geological structures that seem very Martian. A volcanic environment is another likely place to go to search for morphological evidence of life on Mars. Hot springs from volcanic areas might have been places where life could have survived when most of the water on the planet had become ice. Springs are a good place to search for stromatolites in volcanic calderas. Impact craters, too, would be expected to be hot areas for tens of thousands of years after impact and might represent other good areas to search, especially the mud-splashed type of base surge crater.

Another category of evidence is that of carbonates which were accumulated on Earth by microbial action. However, examples of such terrestrial rocks that are only one million years old show no fossils. While carbonates may indicate that there was life on Mars, we are not likely to find any fossil evidence of that life form within the rocks. However, the presence of carbonates might be accepted as evidence of microbial life having fixed carbon from the atmosphere into sediments of ancient bodies of water.

The next chapters examine the various missions being proposed for a return to the Red Planet perhaps to settle the question of the "missing link" between living things and the prebiotic molecules which we believe were present on the surfaces of the terrestrial planets soon after their formation.

4

THE SPIRIT OF 76 YIELDS TO SOVIET ARMADA?

Both the U.S. and the USSR possess the technological knowledge for advanced missions, unmanned and manned, to Mars. As for these capabilities to explore Mars, Soviet space technology is as good as any technology can be with consistency, although the Soviets appear to have had extraordinarily bad luck in all their Martian missions to date, despite their successes with Venus missions. Over twenty-eight years 17 of the 19 Soviet Mars missions failed.

US space technology suffers from an "on-again, off-again" basis of funding, with annual fights in Congress to retain even approved programs. Currently in space we have had over the last decade a "star wars" technology emphasis, whereas the Soviets have concentrated more on a peace technology. We are also limited in launch vehicles because we made the serious mistake of concentrating all our efforts to develop a single launch vehicle—the Space Shuttle transportation system. While the Space Shuttle has been effective in launching military satellites and is essential for the Strategic Defense Initiative, it was only in 1989 that it was used to launch planetary missions—*Magellan* to Venus and *Galileo* to Jupiter.

It normally takes the U.S. ten years from starting a mission to the actual flight. Nothing was planned by the U.S. for starts of interplanetary programs during the doldrums of the 1980s Star Wars Maginot mentality. We do not seem to have learned the lesson of history that there is no ultimate physical defense against any weapon system. The only defense is mental; elimination of weapons by international trust, compromise, tolerance, and understanding of alternate viewpoints. Expending enormous human effort and astronomical amounts of money on defense systems has inevitably proved to be a wasted effort. A determined enemy, be it nation, terrorist, or assassin, can always circumvent any material defensive system.

At this stage of their Mars program, which extends into the next century, the Soviets are looking for conditions on Mars, not for life. Nevertheless, the Soviets are moving to check if there are oases on Mars where life can still be present. They intend to survey the planet in sufficient detail to plan a logical search. It is generally expected that the space missions of the USSR to Mars will ultimately work very effectively; their space missions always have been highly successful when a suitable spacecraft has been developed. The Soviets are also taking over world leadership in space by working closely with other nations of the third world and of Europe. Go into homes in developing countries and you see photos of the cosmonauts from those countries hanging proudly on the wall next to their political leaders. The Soviets have had an excellent international public relations campaign by encouraging cosmonauts from many countries to spend time on the Soviet space stations. By contrast the US has not had a consistent space program and has often offended other nations by cancellations or delays.

The Life Science Division of NASA developed in the mid-1980s a vigorous program of international cooperation to overcome the limited access to space for life science experiments by consolidating worldwide research talent in the space life sciences. Formal joint working groups were formed with the European Space Agency (ESA), Japan (NASDA), the Soviet Union, France (CNES), and the Federal Republic of Germany (DFVLR). The space life sciences community is still relatively small and widely distributed but a number of international projects have been initiated in recent years particularly in regard to developing life sciences experiments and facilities for use in the Space Station program. Ultimately all will be important steps to preparing for manned missions to Mars and the establishment of bases on the Red Planet.

Unfortunately, however, officially the U.S. is still not ready to do bio-experiments on Mars even though these have been proposed by a number of exobiologists and other scientists. Many times the Soviets have offered cooperation with the U.S., and, indeed, a major part of the U.S. biosatellite research into biology under conditions of microgravity has been undertaken on Soviet Cosmos satellites because the U.S. has suffered from a paucity of such spacecraft. It is important to note that cooperation between the U.S. and USSR in space does not require technology transfer, which appears to be a paranoid fear of several U.S. administrations and Congresses. It is also important to note that Congressional fears actually preceded the Reagan administration and may be expected to continue irrespective of the presidency.

The successful fully automated launch, orbiting, and landing of the Soviet space shuttle demonstrates an emerging high level of technology, equal if not superior to that of the U.S. Moreover, the Soviet big boosters have no equal in the U.S. or elsewhere. In recent years, too, the sophistication of Soviet science experiments for exploring the terrestrial planets has leapt ahead, leaving the U.S. leading only in the exploration of the outer Solar System with the *Galileo* and *Cassini* missions that are unfortunately over a decade behind the *Voyager* mission to the outer planets.

After the phenomenal success of the *Viking* mission to Mars, the U.S. space program virtually abandoned Mars as a lifeless world. It was, however, quite obvious that the Soviets, despite their problems with early unmanned missions to the Red planet, had a long range plan to return to Mars. Reports filtered in from Russia about the development of large boosters equal to the *Saturn V* that was used to send U.S. astronauts to the Moon, of plans to keep cosmonauts in space stations for periods of one year, of Russians established on Mars by the centennial of the Soviet revolution.

It became quite apparent that the Soviets were fully aware of the possibilities offered by expansion into space and exploration and development of other worlds. A logical extension of the aggressive space program from the first *Sputniks* into the 1980s was to assure a permanent Soviet presence in space, and to develop cost effective ways to tap the almost unlimited economic resources of the Solar System. By the mid '80s the Soviets had made twice as many successful space launches as the U.S., they had amassed many more hours in space than U.S. astronauts, and they had a much greater launch capability, a permanent presence in Earth orbit, and a well-defined plan for the future of their program.

The extended tours of duty in the Soviet space stations had an obvious purpose—to gain experience for a manned mission to Mars. Since crews can be easily moved to and from stations in Earth orbit, there is no need to push for long terms of duty in a space station. On a manned voyage to Mars, by contrast, crews must be able to function in the space environment for many months, must live away from Earth for as long as two years, and must be able to function for long periods on another world with a different gravity. This type of experience can be gained in Earth orbit, and it is precisely what the Soviets were doing with their cosmonauts in their space stations. They were readying for an assault on the Red Planet, beginning with a new group of unmanned spacecraft to be followed relatively quickly by interplanetary machines comprising a group of spacecraft capable of moving people and freight between worlds.

A fascinating period of human expansion and challenge was about to begin, greater in importance than the expansion of the Europeans to the far-flung continents of Earth and probably equal in importance, if not more important, to that period when life moved from the ancient oceans of Earth onto the dry land.

The Space Research Institute of the USSR Academy of Sciences has proposed a large-scale exploration of Mars as one of the most important aspects of the Soviet space program over the next twenty-five years. Following many years of success with spacecraft orbiting and descending to the surface of Venus, the Soviets have turned their attention from that inhospitable planet to solve some of the mysteries of Mars. They seem to have more interest in searching for extant life on Mars than American scientists do, but less interest in exobiology as a science. The plan appears to be initially to concentrate on global studies of the planet's surface and its atmosphere, and on the return of soil samples. This activity is in preparation for a manned mission early in the twenty-first century as the next logical step in exploring the Red Planet and the beginning of human expansion beyond the Earth, fulfilling the dream of Konstantin E. Tsiolkovsky when he wrote about rocket propulsion and spaceflight nearly a century ago: "perhaps to found a settlement beyond the limits of the Earth's atmosphere . . . the means whereby man can conquer other worlds."

Top level Soviet scientists have described the Soviet program, in great detail at international conferences such as the Lunar and Planetary Science Conferences and the International Astronautical Federation Congresses, leading to an assurance that the Soviet Mars program is real and that funding is planned and not merely hoped for, as is the case with many advanced U.S. programs. The Soviets do not usually make announcements of projects unless they are sure that the projects will be attempted.

The first in a new series of Soviet Mars missions after 14 years of ignoring Mars was

FIGURE 4.1: Diagram to illustrate the configuration of the *Phobos*, a new breed of Soviet spacecraft developed from the Halley *Vega* spacecraft as a means to mount a new series of expeditions to Mars. Unfortunately the first two missions to Phobos with this spacecraft failed, although *Phobos 2* did obtain images of Mars and of Phobos.

the *Phobos* Mission to a Martian satellite in which more than a dozen countries cooperated with the Soviets. The U.S. and the United Kingdom were not involved in experiments but were to assist in analyzing the returned data. Two three-axis stabilized spacecraft named *Phobos* (figure 4.1), launched on July 7 and 12, 1988, heralded a new Soviet assault on the Red Planet as the next stage in Russian exploration of the planets. The Phobos mission was a logical extension of the technology developed by the Soviets for the highly successful Halley *Vega* mission, including the use of international science teams. While earlier Soviet Mars spacecraft had not been successful, the *Vega* was an undoubted high technology success. It provided a spacecraft that was readily adaptable to a new attempt at exploring the increasingly intriguing planet Mars.

Unfortunately, one of the new Mars-bound spacecraft failed soon after entering the transfer orbit from Earth to Mars. An incorrect command sent to the spacecraft in August 1988 had effectively shut down the spacecraft. Its antenna turned away from Earth and prevented further communication. The problem had been compounded because the solar panels could no longer generate electrical power from solar radiation and the spacecraft quickly became a power-starved derelict drifting forlornly toward Mars. It passed by Mars in January 1989 without firing its retrorockets.

The remaining *Phobos* spacecraft took about 200 days to coast to Mars, but began experiencing trouble with some of its electronic systems in December 1988. When it arrived at Mars early in 1989, rocket motors fired on January 29, 1989, to place the spacecraft in an orbit inclined one degree to the planet's equatorial plane, with a periapsis of 500 miles (800 km) and an apoapsis of 49,000 miles (79,000 km), and a period of 76.5 hours. After a relatively short period of several days in orbit, the spacecraft had its periapsis raised to 6000 miles (9700 km), and the apoapsis was lowered to the same height to achieve a circular orbit with an orbital period of about eight hours. The main thrusters were jettisoned getting ready for the spacecraft to approach Phobos by using smaller thrusters.

Phobos has a mean diameter of 13 miles (21 km); its shape is elliptical but quite irregular with several large impact craters (figure 4.2). The largest of these, Stickney, is

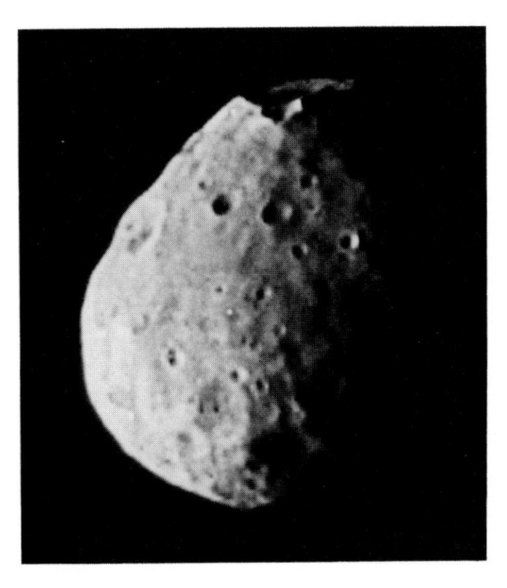

FIGURE 4.2: A general view of the Martian satellite Phobos photographed by *Viking 1* in 1976. North is at the top of the picture while the part of Phobos which always is turned toward Mars is at the lower left part of the satellite. The large crater near the north pole is Stickney which is about 3 miles (5 km) in diameter. About half of the satellite facing the camera is in darkness in this view. (NASA/JPL)

6 miles (10 km) in diameter and was almost large enough to break the small satellite into two pieces. There are two other craters, Hall and Roche, half the size of Stickney, and many smaller craters, a number of which have been subdued by erosion or by a covering of debris. Phobos is 5827 miles (9378 km) from the center of Mars, which places it only 3718 miles (5984 km) above the mean surface of the planet. It revolves around Mars in 7 hours 39 minutes, and because it is tidally locked to the planet it rotates at the same rate, keeping one face turned toward Mars at all times. The escape velocity from Phobos' surface is 49 feet per second (15 m/s) which is about 33 miles per hour. Even with this low escape velocity, material ejected from crater producing impacts appears to have been retained by Phobos to create a regolith that has been estimated to be as deep as 350 feet (100 m).

An unusual feature of Phobos is a system of grooves (figure 4.3) marking most of its surface, entirely different from the other satellite, Deimos. They appear to be connected with the large crater Stickney from which they radiate. They look much like layers of internal structure revealed at the surface, possibly they are regolith-filled fractures which resulted from the impact that blasted Stickney out of the satellite. Their true nature is, however, still a mystery.

Phobos, and Deimos for that matter, are believed to consist of similar materials to those found in carbonaceous chondrite meteorites. These meteorites are richly endowed with organics and water similar, it is believed, to certain asteroids. If so, these satellites will be useful natural resources for the human expansion to Mars since their materials might be tapped for rocket propellants.

The spacecraft obtained images of Phobos when between 530 miles (860 km) and 685 miles (1100 km) from the satellite on February 11, 1989. One image showed Phobos against the background of Mars; others showed closeups of the satellite. Forty images were obtained at distances from 250 miles (400 km) to 125 miles (200 km). All images were precisely positioned on the frames showing that the navigation and orien-

a, left
b, right

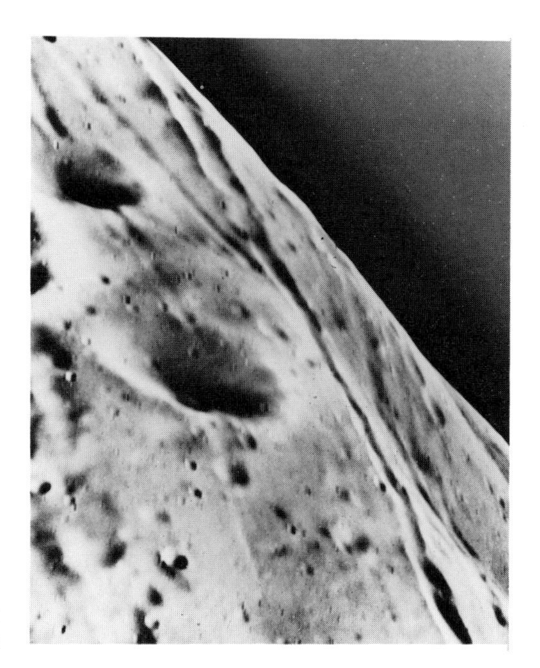

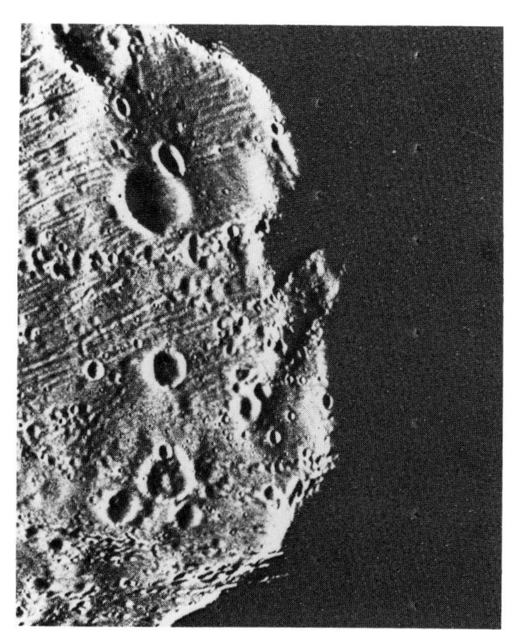

FIGURE 4.3: a) This close-up image of the surface of Phobos covers an area of 1.86 by 2.17 miles (3 by 3.5 km) and shows details to a size of about 50 feet (15 m). The picture shows a region in the satellite's northern hemisphere which has many craters and striations. The striations are about 300 to 700 feet (100 to 200 m) wide. (NASA/JPL)
b) The striations are shown in greater detail in this view. Their origin is unknown. They might have been formed by ejecta from craters on Phobos, by a cloud of debris colliding with the satellite, or a fracturing of the satellite's material when it was part of a larger body. The Soviet mission to Phobos may have thrown more light on these mysterious features had it been successful. (NASA/JPL)

tation of the spacecraft had been highly successful. The spacecraft also obtained images and other data concerning Mars, including detailed heat emissions along an equatorial band, possible evidence for Martian radiation belts, and leakage of the Martian atmosphere into space.

The spacecraft gradually moved closer to Phobos until by early in March the satellite and the spacecraft were separated by only 60 miles (100 km). Controllers aimed to have the spacecraft hover over the satellite by mid-April. But before it could move closer to Phobos the antenna directing system malfunctioned and communication with Earth was lost irrecoverably on March 27, 1989. The spacecraft had earlier encountered some trouble with its systems and was operating on its backup transmitter and only one battery. Some US observers attributed the failure to a computer malfunction compounded by a delay in communications with the spacecraft because of priorities to communicate with a Soviet space station.

The intention had been to use short burns of the rocket engine to send the Russian spacecraft to intercept the satellite. If the spacecraft had approached to about 165 feet (50 m) from the satellite in April and May 1989 it would have deployed two small spacecraft intended to land on Phobos. The main spacecraft would have remained somewhat further away from the satellite for 140 days while the landers conducted their initial mission; one hopping about the small satellite, the other tethering itself to the satellite. Unfortunately, the loss of communication with the main spacecraft made these lander missions impossible.

A Long-Term Automated Lander carried by the main spacecraft was designed to make a controlled landing on the surface of Phobos and continue to operate there for about twelve months. It carried an alpha-backscatter and X-ray fluorescence spectrometer (Germany), a sun-sensor (France), a penetrator probe (USSR), accelerometers and temperature sensors, (USSR), a TV camera (France and USSR), a seismometer (USSR), and several radio systems (USSR and France). Solar panels were intended to be unfurled after landing and oriented toward the Sun by an optical sensor. They were mounted on an instrumented platform which was held on legs several feet above the surface. The solar sensor was to be used to record librations of the satellite as it moved around its orbit.

The Hopping Lander was intended to move about the surface on up to ten mechanically induced hops, each covering about 60 feet (20 m). Between hops, when on the surface, this lander would have oriented itself, made scientific measurements, and relayed data to Earth. The instruments carried were an accelerometer, X-ray spectrofluorimeter, a penetrator, and a magnetometer, all by USSR scientists.

It was intended that the mission could be extended for about a year to provide important information on the chemical, thermal, physical, and magnetic properties of the rocky surface of the satellite. Such information is important to the Soviet long range program of Mars exploration because the satellite might be useful as a staging station or as a refueling and materials supply station for manned expeditions to the Red Planet, as first suggested in the author's book *Rocket Propulsion* published in 1952 (figure 4.4).

Instruments aboard the main spacecraft included an innovative laser mass spectrometer (LIMA-D; Austria, Bulgaria, Germany, Finland, and USSR) designed to fire a laser beam to vaporize materials on the surface of the satellite and to analyze the scattered ions, and a remote mass analyzer to generate ions by firing an ion beam at the satellite (DION; Austria, France, Finland, and USSR). A radio sounder (GRUNT; USSR) on the spacecraft was to scan the surface of Phobos at three resolutions. Three stereo TV cameras (Bulgaria, Germany, and USSR) could store 1100 frames and resolve surface details down to 2.5 inches (6 cm). When used to observe Mars, these cameras were designed to resolve Martian surface detail down to 4.3 miles (7 km).

An infrared spectrometer and radiometer system (KRFM-ISM-TERMOSCAN; USSR) was intended to provide thermal maps of Mars and of Phobos and to search for thermally hot areas and areas of permafrost on the planet. A gamma-ray spectrometer (GS-14; USSR and France) was to be used to help ascertain the chemical composition of the rocks of Phobos and Mars by measuring gamma emissions produced by natural radioactive decay of uranium, thorium, and potassium, and the interaction of cosmic rays with other elements. This complex array of instruments was lost when communication failed.

Directed solely to Martian exploration were instruments to search for water on the planet, and to analyze the Martian atmosphere seeking details of its ozone, water vapor, dust, oxygen, and carbon dioxide concentrations, and recording any seasonal changes. These used six sets of neutron counters (IPNM; USSR) to measure neutron flux intensities over several energy ranges to determine the bound water content of the surface layers.

The *Phobos* spacecraft carried a commemorative plaque intended to be placed on the Martian satellite. The Soviet Space Research Institute and the Soviet government agreed to a suggestion by NASA's Solar System Exploration Office that the spacecraft should carry an aluminum plate with a photographic transfer of the U.S. Naval Obser-

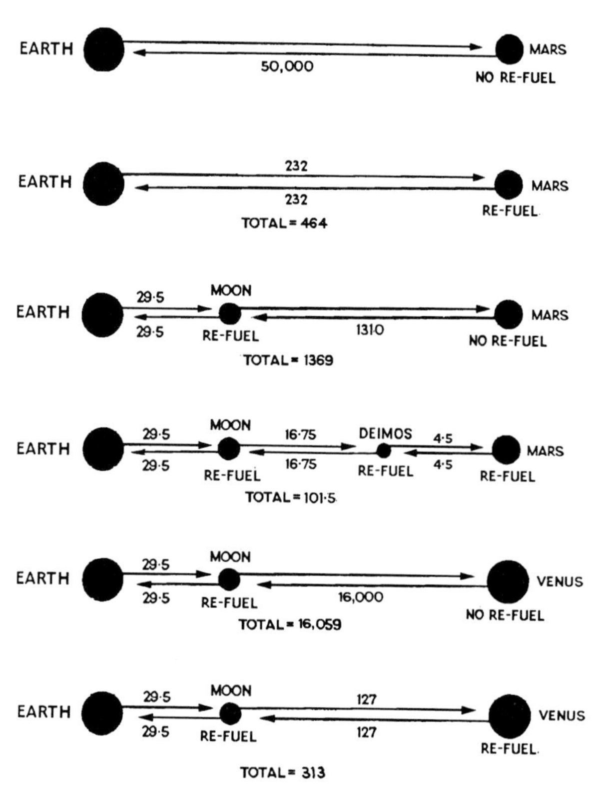

FIGURE 4.4: This illustration from *Rocket Propulsion* by the author, published by Chapman and Hall in 1952, shows mass ratios calculated for various interplanetary journeys. It illustrated the tremendous gains that could be made by establishing refueling stations on other worlds and showed that a voyage to Mars could become feasible if refueling were possible on one of the Martian satellites, in this case Deimos. The Moon was also suggested as another refueling station for interplanetary journeys.

vatory's telescope logbook page dated August 17, 1877. That was the date Asaph Hall recorded the position angles and separations of Phobos and Deimos as satellites of Mars. He had first observed Phobos on the night of August 11 and referred to it as the Mars star. On August 17 he spotted Deimos. On the log he wrote, "both the above objects faint but distinctly seen by both G. Anderson [his assistant] and myself." The public announcement was made by the observatory on August 20.

The plaque, in addition to reproducing Hall's original notations, carried the following two citations, in both English and Russian: "USSR 'Phobos' Mission 1988" and "Discovery of Phobos—Asaph Hall—U.S. Naval Observatory—August 17, 1877."

The *Phobos* mission was to have been a great step forward in the exploration of Mars, comparable with the advance to the *Vikings* from the earlier unmanned missions to the planet. In addition, the *Phobos* mission was to demonstrate rendezvous and control technologies needed for a mission to bring back a Martian sample and the subsequent manned mission which the Soviets plan to send to the Red Planet. Rendezvous technologies were, indeed, demonstrated. But when the spacecraft failed it was a serious blow to Soviet exploration of Mars.

Despite the undoubted setback to the Soviet program, the *Phobos* spacecraft provided important new information about Mars and its satellite, Phobos. This information is extremely valuable for the planning of future missions. The thermal mapping with TERMOSCAN provided very detailed images of a 950-mile (1500-km) wide swathe of

Mars along the equator with higher contrast and resolution than previous images obtained by television cameras at the same orbital altitude of about 3700 miles (600 km). The images show that outflow channels (ancient "river" beds) and areas of chaotic terrain from which it is believed the floods originated, are cooler than surrounding terrain.

Another equatorial scan by the ISM instrument provided data on the amount of hydrated minerals in the equatorial regions. The results appear to confirm that there are sedimentary deposits on Mars. This is important to seeking sites at which subsequent missions should look for evidence of extinct or extant life.

Phobos carried important instruments to explore the interaction of the solar wind with Mars, experiments which had been omitted from previous missions to the Red Planet. A turbulent magnetic tail was found extending away from the sun. It was discovered that the solar wind is stripping Mars of atmosphere at the rate of nearly 90 tons per day (86,500 kg), mainly in the form of ionized oxygen. This would suggest that if Mars did not have a stronger magnetic field in the past to protect it from the solar wind as Earth is protected by its strong field, an atmosphere equivalent to the terrestrial atmosphere could have been lost into space during the lifetime of the planet. A planetwide ocean some 33 feet (10 m) deep could have been lost by this process, which might explain where the water responsible for the great floods has gone. Fortunately for a permanent human presence on Mars, the planet appears to still have water trapped in hydrated minerals and in the polar caps and what is believed to be an icy regolith and deep permafrost at high latitudes. The presence of hydrated minerals on the slopes of the big Tharsis volcanoes suggests that these volcanoes have been outgassing water. Moreover, without some means to replenish it, today's Martian atmosphere could be lost to the planet in 100 million years. The atmosphere is most likely now being replenished by material from the polar caps. *Phobos* also mapped fine structure in the Martian magnetosphere.

Observations of Phobos determined that there is much less water in the surface materials than had been anticipated on the basis of ground-based observations. However, this does not mean that the interior of the satellite is also very dry. If it were so that might cast doubts on the current belief that the Martian satellites are captured asteroids of a carbonaceous chondritic composition. Although the spacecraft confirmed that the surface of Phobos is a universally dull gray color, the spacecraft's instruments discovered that the surface is not homogeneous in its composition.

Observations from the spacecraft were also extremely valuable in obtaining a more precise definition of the position of the satellite in its decaying orbit. For future missions the location of Phobos can now be predicted one hundred times more accurately than previously; to within 2.5 miles (4 km).

The next stage of the Soviet Mars exploration program is expected to start in the 1990s with measurements to be made from an orbiter, in the atmosphere by balloon, and on the surface by a roving vehicle. However, there are doubts that a rover will be launched much before the end of the decade. If the technology can be developed of assembling the Mars spacecraft in Earth orbit and using aerobraking rather than retrorockets to enter an orbit around Mars, the Soviets state that payloads carried to Mars can be increased significantly. A payload of 3300 pounds (1500 kg) could become commonplace, they say.

The Soviet Mars program is ambitious and well planned, with a sample lander expected to be on Mars by the mid 1990s. But the USSR has no strong exobiology programs for Mars. The Soviets have stated that after the mission to Phobos they will undertake the following programs of missions to the Red Planet. The failure of the mission to Phobos may, however, produce a slippage in these program schedules.

In the period 1990–1996:

1. An orbiter in polar orbit carrying instruments for remote sensing to make a global survey of Mars.
2. A balloon deployed in the Martian atmosphere for studies of the atmosphere and the planet's surface.
3. A small rover soft-landed on the surface for initial exploration and proof of technology.
4. A network of small stations landed on the surface for global meteorological studies.
5. A subsatellite for mapping the gravitational field of the planet and its anomalies and their relationship to topographic features.
6. A device to return a container with photo film of the Martian surface taken with a super-high-resolution camera to test the technology of cargo return from Mars and to provide in a reasonable time high-resolution maps of much of the surface of Mars.

In the period 1996–2010:

7. A Mars sample-return mission in which a rover will pick up samples of the surface of Mars, return them to a spacecraft orbiting Mars which, in turn, will send the samples back to Earth for analysis.

In the period 2010–2025:

8. Manned spacecraft to Mars with landing of cosmonauts on the surface, with possible development of refueling stations on Deimos or Phobos.

In the period 2025 onward:

9. Manned space systems for transportation between Earth and Mars, outposts, and bases, research stations and laboratories on Mars, and possibly on its two satellites, for a permanent Soviet presence on Mars. This will reflect the Soviet experience in Earth orbit where they have had a permanent presence in space stations there while the U.S. continued to make breathless little forays into the space environment.

As with the U.S., USSR space scientists are concerned about contamination of Mars by terrestrial biology, and all the Soviet spacecraft are sterilized to prevent this from occurring before a thorough search has been made for life on Mars and cosmonauts are landed there. In fact, there is an international agreement aimed at preventing contamination of other worlds by terrestrial biology, an agreement that has been scrupulously followed by the U.S. and the USSR.

The ambitious Mars missions of 1992 and 1994 consist of several important elements for exploring Mars. There were to be two separate missions using a *Phobos* type spacecraft; a Mars 1992 orbiter with a balloon, and a *Mars-94* mission with a lander and a rover. However, these missions have now slipped to 1994 and 1996 with the rover in the 1996 or later mission. The failure of the two *Phobos* spacecraft will undoubtedly be reflected in Soviet planning, as will attempts to revive the flagging Soviet economy. However, as with the U.S., space activities can produce economic multipliers which are lacking in military expenditures.

Exact configurations of Soviet missions are usually finalized just before launch, a flexibility made possible by their tremendously powerful launch capability.

The main element of the first mission of the 1990s is an orbital station, an artificial satellite of Mars, carrying a scientific payload of 880 pounds (400 kg), including a system for taking high-resolution pictures of the Martian surface. At one time it was planned that a return rocket would be carried and used to bring photographic materials weighing 66 to 88 pounds (30–40 kg) from the Martian orbit back to Earth. This would weigh some 660 pounds (300 kg). This film return has now been shelved, at least for the time being. The orbital station may also deploy a balloon station, penetrators, and meteobeacons. The penetrators are not yet available and the Soviets hoped they might be provided by the US which has done much theoretical work on penetrators.

A lander, which may be delayed to the 1996 or later launch window, will carry a rover and a balloon station, and will weigh 1100 pounds (500 kg). It will also carry two penetrators each of 220 pounds (100 kg), and a subsatellite of 110 pounds (50 kg). Several meteobeacons for hard landings will also be carried. The Soviets have stated that they plan a rover sample gathering and return mission with launches in 1996 and 1998 following the 1994 mission.

Two of the advanced *Mars-94/96* spacecraft will be launched in 1994 and two more in 1996 to follow an Earth-to-Mars trajectory taking about 300 days to reach the Red Planet. Each spacecraft will use the atmosphere of Mars to slow its path so that it enters an orbit around the planet. Aerocapture uses the drag of the Martian atmosphere to reduce considerably the amount of propellant that would otherwise have to be carried to accomplish the mission with a given payload weight. It allows a more complex mission to be planned and executed with a launch vehicle of a given capability.

To achieve aerocapture a precise navigation system is essential. Angular coordinates of the spacecraft have to be determined from the ground by using radio interferometry. As each spacecraft approaches its encounter with Mars it will be guided with the help of optical TV-type instruments. These provide referencing of the planet and the spacecraft relative to stars with sufficient precision to allow the pericenter of the hyperbolic approach trajectory to be adjusted within 6 to 13 miles (10–20 km). This accuracy makes it possible, claim the Soviets, for the spacecraft to use aerodynamic braking effectively to enter an elliptical orbit around Mars which can then be adjusted later by rocket thrust. Next, the spacecraft corrects its orbit from the initial ellipse to the more circular observation orbit from which prospective sites for the rovers, balloon stations, meteobeacons, and penetrators are reconnoitered and evaluated.

When the spacecraft of the Soviet lander mission have successfully entered orbit, each will release its landing vehicle which then uses rocket braking to slow down out of orbit and descend to the surface, deploying a balloon station to the atmosphere on the way down. When each lander is safely on the surface, its rover vehicle trundles down a runway and is ready to begin its exploratory mission. Next, the orbiter releases ten meteobeacon cassettes for hard landings, and despatches two penetrators to impact and penetrate the Martian surface at two chosen sites. A subsatellite is also released and sent into a suitable orbit for it to map the gravitational field of Mars.

The research program continues gathering data by the orbiter, the rover, the balloon, the meteobeacons, the penetrators, and the subsatellite. Finally, when all the photomaterial has been gathered within the orbiter, a capsule containing the photographic film is released and rocketed back to Earth.

The objectives of the Soviet unmanned Mars missions cover several main areas of science: studies of the surface, studies of the atmosphere, monitoring the Martian climate, studies of the interior of the planet, and studies of plasma and astrophysics.

For the surface study, the Soviets intend to make a topographic survey of the surface of Mars including high resolution measurements of the terrain. An aim is to produce maps of the distribution of minerals on the surface, study the elemental composition of the regolith, explore the zone of permafrost and its structure, and search for organics in the soil and the subsurface materials.

Concerning the atmosphere and climate, the objectives are to determine the minor constituents of the atmosphere, including water vapor, carbon monoxide, and oxygen, and how they vary with time and with altitude above the surface. Instruments will search for areas of high humidity, determine how temperature varies in different regions of the atmosphere, and record pressure variations. Typical features of the composition of the atmosphere near volcanic mountains, such as Olympus Mons, will be checked for current volcanic emissions that might signify that the volcanos are still active. The characteristics of aerosols will be sought, and the neutral and ion composition of the upper atmosphere will be measured.

As for the planet's internal structure, the Soviets say they will attempt to ascertain the thickness of the planet's crust and the dimensions of its fluid core. The magnetic field will be mapped and heat flux from the interior measured.

The interaction of the planet with the solar wind will be studied to determine the parameters of Mars' magnetic dipole, its moment, and its orientation. Experiments will attempt to map the three-dimensional distribution of ions and energy within the plasma surrounding the planet and will record plasma waves in the electric and magnetic fields. The structure of the magnetosphere and the extent of its boundaries will be mapped. During the interplanetary journey to Mars the spacecraft will also record solar wind fluctuations and will search for cosmic gamma-ray bursts.

The Soviet scientists have suggested cooperation with the United States during this program in the form of joint research with the American *Mars Observer* mission planned for about the same time. There are many opportunities for cooperation. Interesting areas of the planet might be investigated by spacecraft of both nations to obtain confirmatory information, including observations by the U.S. spacecraft of the surface areas where the Soviet landers, rover, penetrators, and meteostations are operating. The U.S. spacecraft might also be used to relay data from the Soviet balloons when these cannot be in contact with the Soviet orbiter because of inappropriate geometry. Back on Earth the ground data receiving systems might be integrated to provide a 24-hour capability of receiving data from either mission. Furthermore, the Russian scientists suggest a joint effort to interpret all the data returned by the Soviet and U.S. missions and to work together to develop an engineering model of Mars that can be used for planning future exploration of the planet.

The Russian Mars-orbiter spacecraft will be placed in a nearly circular orbit with a periapsis between 125 and 310 miles (200 and 500 km) above the Martian surface. The orbit will be highly inclined to permit coverage of the intriguing polar regions. A preliminary selection of instruments to be carried in addition to the TV cameras include a stereo TV surveying camera system and a TV high resolution (about 65 feet or 20 m at the surface) system with a video spectrometer. There may also be a super high resolution (about 3 feet or 1 m at the surface) camera system that will store its images on photographic film for return to Earth.

Only a small part of the surface of Mars can be mapped in detail with drifting balloons and moving rovers. Greater areas can be covered from orbit, but there are limitations on the amount of pictorial information that can be transmitted back to Earth during the mission. To image the Martian surface at 3 feet (1 m) resolution and transmit the data at 100 kilobauds would require 8.6 years to cover only 1 percent of the surface, or 258 years to cover 30 percent of the surface. However, to return the same amount of pictorial information on photographic film carried by a cassette in a return rocket would require that the return rocket carry a payload of only 2.75 lb (1.25 kg) or 82.7 lb (37.5 kg) respectively for these two coverages. Future missions may use such a cassette recovery to transfer data on 30 percent of the Martian surface back to Earth.

A two-channel Fourier spectrometer will cover the range from 1.2 to 40 micrometers. There will be a spectrometer for sounding the atmosphere by occultations of Sun and stars and for recording emissions from the atmosphere. The payload will also include an infrared radiometer, a helium-line photometer, a long-wave radar, a millimeter range radiometer to record the lines of water and carbon dioxide, a gamma-ray spectrometer, a neutron spectrometer, and a neutral mass spectrometer. The Orbiter will also carry a magnetometer, a package of intruments to measure plasma, a dust-particle counter, and instruments to study stellar oscillations and cosmic gamma-ray bursts. Scientists of many nations are involved cooperatively in these experiments.

The envelope and gas carried by the balloon station amounts to 55 to 120 lb (25–30 kg), and the gondola carrying the payload is 44 lb (20 kg). The balloon is expected to drift for 10 to 15 days and cover up to about 1250 miles (2000 km) of the surface. It will transmit 50 megabits of data per day for relay via the orbiter back to Earth.

Jacques Blamont of the French Space Agency (CNES) who has worked previously on balloons to float in the atmosphere of Venus, proposed a double balloon for a Soviet Mars mission. An inner balloon filled with helium provides the stable buoyancy, while a second balloon is open to the atmosphere and relies upon solar heating for buoyancy. This configuration allows the balloon system to descend to the surface at night and rise into the atmosphere during the daytime. It also allows the balloon's altitude to be changed during the daytime by control of venting. This altitude control is achieved without a need for consumables such as helium gas or ballast material. However, recent reports seem to imply that a simpler, single balloon concept will be used, at least on the early missions.

The scientific payload amounting to less than 22 lb (10 kg) consists of a panoramic TV camera with a surface resolution of about 3 feet (1 m), a surveying camera with a surface resolution of 4 inches (0.1 m) from 600 feet (200 m) altitude, and a system of meteorological instruments to measure pressure, temperature, humidity, wind velocity, and aerosol density. In addition, the balloon's payload will include an instrument to sound electromagnetically the layers of regolith underlying the surface and to determine the depth of permafrost. Another instrument will be used to analyze differential temperatures of and volatiles in the soil. There will be spectrometers for analyzing alpha backscattering and X-rays, and a photometer. A magnetometer might be carried to determine the crustal magnetic field of Mars along the path of the balloon.

Balloon systems of this type have shown to be practical in tests carried out on Earth. For example, scientists from California Institute of Technology and the Jet Propulsion Laboratory tested such a balloon system in the flat expanse of the Mojave Desert of Southern California during the winter of 1986. The balloon performed as expected. Only a short period of exposure to weak December sunlight was needed to raise the balloon to several hundred feet above the surface of the flat desert, and when landings were made they were soft enough not to present hazards to a scientific payload. Other successful tests were completed in Europe during 1988.

The balloons will rise in the Martian atmosphere during the daytime and descend toward the surface at night. When first deployed they will descend into the atmosphere by parachute and during this descent they can be programmed to release penetrators. At the appropriate altitude the parachute will release and the balloon be inflated.

The rover vehicle is expected to have a mass of 330 lb (150 kg) and an active lifetime on Mars of up to three years. It is expected to be able to cover a total distance of 60 miles (100 km) during its mission. The orbiter's high resolution imaging system will be used to select the track of the rover on the surface. The scientific payload of about 45 to 80 lb (20–35 kg) will include a TV camera, an instrument to measure soil composition, a gamma-ray spectrometer, a system for analyzing volatiles in the soil, and another system to examine microscopically the fine structure of the soil. There will be meteorological instruments, systems to sound the surface electromagnetically, and seismically, a core sampler, and a system to search for traces of biological activity. The soil will be core sampled to a depth of several yards. Much of the technology required for operating a Martian rover was developed by the Russians many years ago for the unmanned exploration of the Moon using roving vehicles on the lunar surface and the return of lunar samples.

An ambitious part of the Soviet unmanned Mars missions is to establish a global network of automated meteorological stations on Mars to measure the planet's meteorology on a continuing basis for at least a year. Ten of these stations are anticipated as forming the first network. If this system operates successfully, the intent is to add more stations by later missions. Such stations offer a number of advantages in exploring Mars. They provide global coverage and can be deployed at sites selected in the most interesting regions such as old river beds, canyon floors, and volcanic caldera, all of which are extremely difficult to access by other means.

Each station will measure pressure and temperature of the atmosphere, surface, and subsurface, and wind velocity. In addition, the stations can be instructed to carry out specific experiments such as measurements of water vapor, measurements of atmospheric transmittance, backscattering by dust particles, intensity of solar radiation at the surface, seismic activity, and magnetic field measurements. The station will be released about three days before the main spacecraft enters Martian orbit. Each station will carry a payload of up to one kilogram and will be intended to operate on the surface for up to two years.

The stations will each operate independently and automatically, their transmitters being switched on several times each day for up to a minute to transmit the data gathered either to an Orbiter or directly to Earth.

The Soviets would like to carry two penetrators on each Mars spacecraft. These are essentially small missiles which plunge dartlike into the Martian surface with a payload of instruments capable to resisting the shock of impact. They are intended to study the soil and the internal structure of the planet. Several penetrators can form a network to support long-term seismic observations. The active lifetime is expected to be up to two years, the scientific payload about 9 lb (4 kg), and the maximum depth of penetration about 16 feet (5 m). The penetrators are about 8 inches (20 cm) in diameter and will impact the surface at about 330 ft/sec (100 m/sec). Each will have an active lifetime of up to two years. The scientific payload of each penetrator will include a seismometer, an instrument to measure composition of the soil, and another instrument to measure heat flux from the interior. Also included will be a system of meteorological instruments and an instrument to measure volatiles in the soil. A TV camera may be carried also.

A subsatellite of about 110 lb (50 kg) mass was included in the ill-fated *Phobos* spacecraft. This small spacecraft was intended to be used to clarify the gravity field of Mars. In a number of future missions the Soviets say they plan to use similar subsatellites to continue research on the gravity field of the planet. In the Mars 1994–96 missions the subsatellite to be released from the orbiter will also weigh about 110 lb (50 kg). Precise measurements of its trajectory will be used to develop a model of the gravity field of Mars and its anomalies; a prerequisite for manned expeditions to the planet.

At the time of writing it is not certain whether or not all these spacecraft and experiments will be flown, especially the meteostations and the penetrators.

In 1998 and 2000 the Soviets plan to send their sample return missions to Mars. The sample return mission is a very difficult mission in its selecting sites and returning samples. The scientific objectives are to study the surface topography, map mineral deposits, determine composition of the soil, investigate the location and extent of the cryolithic zone, and check for organics in the soil. The mission will also study the atmosphere and its composition; investigate the inner structure of the planet and the plasma surrounding the planet and its interactions with the solar wind.

The Mars Sample Return mission may be sent to Mars by means of the Proton SL-12 launch vehicle according to Phillip Clark writing in *Spaceflight,* a publication of the British Interplanetary Society. This launch vehicle can place about 25 tons in low Earth orbit and would permit sending about 6 tons of payload to Mars.

If the launch takes place in 1998 after the 1994 and 1996 orbiter and rover missions it is expected that an armada of four spacecraft will be despatched to the Red Planet. Two of these Russian spacecraft will be orbiters and Earth-return complexes, and the other two will be for landing on Mars and return to Mars orbit. As is usual with Soviet missions, as opposed to U.S. missions, spacecraft are launched in pairs for backup by redundancy. The last time this was done in a US mission was the Voyager mission to the outer planets, launched in 1977.

The Soviet orbital station for the sample return mission weighs 880 lb (400 kg); the return vehicle for the photographic and other material is 660 lb (300 kg). The lander which will carry the roving vehicle weights 1100 lb (500 kg), and several penetrators each weigh 440 lb (200 kg).

After an interplanetary trip of about 300 days, the orbiters will arrive at Mars and probably be aerobraked into an elliptical orbit inclined about 30 degrees to the Martian equator. The chosen landing sites will be mapped in detail once again, and alternatives found should they not prove as suitable as expected.

Within a month after the orbiters reach Mars the landers will arrive. Each will separate from its interplanetary bus which will then fly by Mars and continue into space, another common Soviet practice. Like the U.S. *Viking,* the landers will be protected by aeroshells and will slow down by aerodynamic braking, parachutes, and retrorockets in a sequence similar to the landing of the *Vikings.* Samples will be gathered on the surface and placed in a capsule within an ascent vehicle. This will later be launched by rocket from the Martian surface for a rendezvous with the orbiting spacecraft. The return capsule will be transferred to an Earth-return spacecraft attached to the orbiter, but this return spacecraft cannot be despatched back to Earth until the planetary configurations become suitable. This may take up to ten months. Then the return vehicle separates from the orbiter, accelerates to transfer velocity and cruises back to Earth's orbit. Eleven months later it makes a direct entry into Earth's atmosphere, is slowed by atmospheric braking, and ultimately lands by parachute.

This would be the simplest type of sample return program, but concurrently with U.S. thinking on sample return, the Soviets appear to be considering a more sophisticated approach in which a roving vehicle is carried to Mars to collect samples from a number of sites along the path of the vehicle. Selecting a single site for sampling does not make economic sense considering the amount of effort that must be put into sending a sample return mission to the Red Planet. It is most likely, as a consequence, that the Soviets will opt for a mission similar to that proposed, but not yet funded, for the U.S. mission which is described in a later part of this book.

No details have been released of the manned expedition planned by the Soviets. It is highly probable that it will be similar to those proposed by several studies published in the U.S. Assembly of the interplanetary vehicle or vehicles will be achieved in low Earth orbit and associated with a Soviet space station acting as a construction base. Following escape from Earth orbit and an interplanetary cruise the interplanetary vehicle may be placed in orbit around Mars, or may rendezvous with a Martian satellite. Crews will then descend to the surface of the planet using aerodynamic vehicles, and freight will be sent down in separate vehicles whose shells can be used to provide habitats on the surface. An ascent vehicle, consisting of a refurbished lander or a separate ascent module, will later be used for return of personnel to the orbital station and their subsequent return from there to Earth.

Requirements for a manned mission to Mars are fairly well defined as discussed in the next chapter, and the Soviets will probably select a mission profile that will make full use of their powerful launch capabilities. The requirements for a base on Mars are also fairly well understood, and these are discussed in a later chapter in connection with the need to push for a U.S. or international program for the exploration of the Red Planet. The message about Mars is, however, quite clear. If the U.S. does not take an international lead in the peaceful development of space and other worlds, the USSR undoubtedly will. By the centennial of the Soviet revolution the Russians will most probably have established a permanent human presence on Mars that will most likely include, as with the Earth orbital missions, cosmonauts from other countries.

5

WAKE UP AMERICA!

In the United States today there is no shortage of ideas for space exploration. At all levels from young school children to college students, from grass roots of laypersons to graduate scientists and engineers, and from government and industrial employees to many managers, there is excitement about planning an American future in space. What is lacking, and has been lacking, is a continued long-term national leadership.

Following the 1986 report of the President's Commission on Space, a report by the solar System Exploration Committee of the NASA Advisory Council entitled *Planetary Exploration Through the Year 2000,* and a subsequent (1987) report to NASA, *Leadership and America's Future in Space,* by ex-astronaut Sally Ride, the Agency began a series of studies of missions such as the Mars Sample Return mission. The Agency has an Office of Exploration, of which Code Z and Project Pathfinder have several purposes including establishing the technology for unmanned and manned missions to Mars, and how they can be accomplished in view of the economic straits of the nation. While a detailed space policy of exploratory missions has been established, and a timetable suggested, approval by the administration and by Congress had not been obtained at the time of writing. It seems that the ball has at least started to roll in the right direction, but it will probably need continued urging by grass roots groups such as *The Case for Mars* workshops to preserve its momentum in view of the astronomical federal deficit resulting from years of irresponsible "guns before butter" politics.

The research required for an all-out manned assault on Mars includes an understanding of how humans will respond to the environment of space for long periods.

Also, we need to define how automation can be used, for example, for Mars rover and sample return missions which requires that teams of exobiologists, planetologists, and computer scientists work together. How do we make on-board analysis of core samples? How can we select the best sites for exploration? Can we use artificial intelligence within the rover to pick targets of opportunity?

There is need for high level control systems to be designed for use in Mars missions, requiring the merging of expertise in automation and robotic devices as well as in "expert" systems and artificial intelligence systems. "Expert" systems are computer software programs which are under development to try to simulate the expertise of human specialists of various disciplines.

Considerable work needs to be done on determining objectives for scientific exploration of the surface of Mars. These objectives include biology, structure, dust storms, and Earth analogs, requiring a close coupling of other sciences and biology and of exobiology in particular. The search is essentially for microfossil evidence of past life on Mars but we should not ignore the possibility of finding present life on the planet.

Exobiology is aimed at understanding the origin, evolution, and distribution of life in the universe. So far we only are sure that there is life on one planet, the Earth. However, we are aware of the cosmic evolution of biogenic compounds, the evolution of prebiotic molecules in space and their presence on meteorites. Also we are aware of the evolution of terrestrial life, but because Earth's history is chaotic earlier than 3.5 billion years ago, we have no evidence of the change from prebiotic molecules to the first living things able to reproduce and evolve into the complex life forms we know today. A major objective in exobiology is to trace the pathways from the original inorganic universe to the evolution of living things and conscious beings such as ourselves.

In trying to ascertain what happened on Mars, we have to trace the history and location of water on that planet. We have to know the atmospheric composition and distribution and how this has changed over time. We have to know the nature and composition of the surface materials and their evolution. We have to search for extinct and extant life; both are very important to our understanding of biological processes on our own planet and elsewhere in the universe since about half of all known life forms on Earth are now extinct. If Mars was originally much like Earth some 3.5 billion or more years ago and no trace of extinct life is found on Mars, we shall have to radically change our theories about the origin and evolution of living things on our own planet.

The U.S. *Mars Observer* mission has been approved by Congress for a launch in 1992. The Mars Rover Sample Return mission had not yet been approved at the time of writing and was merely a study project. The study suggests a late 1990s launch date as a feasible target. In planetary exploration it is instructive and somewhat disheartening to observe that all the major space missions of the United States in the thirty years of NASA's existence were conceived during the agency's first decade, and mostly as a result of crusading by a relatively few dedicated individuals of the scientific community. One third of NASA's existence, the last decade, saw no new space missions approved or launched. In that period nearly all the space agency's activities were concentrated just above Earth's atmosphere! Fortunately the situation is now changing as NASA's Office of Space Exploration develops detailed plans for human expansion into space.

The most recent deep space interplanetary mission was that of *Pioneer* to Venus in 1978 which recently celebrated ten years in orbit actively observing that planet; a remarkable achievement for a mission costing about $200 million. The *Pioneer* Venus program originated from a group of scientists—R. M. Goody (Harvard University), D. M. Hunten (then of Kitt Peak National Observatory), and N. W. Spencer (Goddard Space Flight Center) who began planning the mission in 1967, eleven years before the launching. The *Pioneer* Venus orbiter accumulated data for the longest period that another planet has been observed continuously. And even in the exploration of Venus the United States could only gain credit for five missions compared with fifteen successful missions of the Soviets to Venus, including successful landings on the scorching surface of that inhospitable world. However, another U.S. mission to Venus, *Magellan,* was sent to Venus in 1989. The United States did much better in successful missions to Mars but not in the number of launches to that planet.

An immediate need is to identify the science instruments required for the scientific exploration of Mars, especially in connection with exobiology. Also, there is a pressing need to identify sites on Mars that are most likely to reveal significant information in this search for evidence of life beyond Earth.

Viking, for example, could not sample much below the highly reactive surface layer of soil. A meaningful search for life on Mars requires that samples be taken from deep beneath the surface and that these samples be protected from the surface environment during the collection process. The Soviets developed a method of doing this on the Moon and took some deep core samples into their lunar roving vehicle.

In the previous chapter it was mentioned that the Russians accepted as an important tool in the investigation of Mars the use of penetrators and hard landers. Dartlike hard landing spacecraft, or penetrators, offer a convenient way to sample at many places on the planet. They also offer the advantage of being able to sample areas which are inaccessible or dangerous for soft landers such as the *Viking,* or for rovers.

As mentioned in the previous chapter, the typical penetrator is a missile which impacts the surface and because of its high speed is able to penetrate as far as 50 feet (15 m) beneath that surface (figure 5.1). Some penetrators may use aerobraking to reduce their speed of impact if necessary for special instrumentation. An afterbody is left at the surface from which information can be telemetered to an orbiting spacecraft. The afterbody can also contain an imaging system to look around the penetrator site and send pictures back to the orbiter. The penetrating part of the missile contains instruments to measure the amounts and the phases of water, the electrical conductivity of the soil, and to conduct inorganic and elemental analyses of the soil. Even a gas chromatograph could be carried since the deceleration on impact, an expected 1000g, does not present problems to modern instrumentation. The necessary technology of shock resistance has been developed for missiles and other weapon systems. The problem is one of carrying sufficient payload, and this is the limiting factor in design of penetrators. A payload of about 15 lb (7 kg) appears to be feasible.

Soft landers, by contrast can carry much larger payloads of instrumentation with capabilities for in-depth analysis of samples. But they may not be able to land safely at locations on Mars which are the best sites to search for biology.

Another major question is whether most effort should be placed on developing a mobile soft lander to conduct all the analysis on Mars, or whether effort should be concentrated on designing a spacecraft that can pick up samples and return them to Earth for analysis here. Certainly it seems that greater analytical capability can always be provided in terrestrial laboratories than on Mars. This is not just a matter of design but a matter of timing also. In planning a Mars mission, U.S. scientists have to start years ahead of the next mission to design the instrumentation within the payload

a

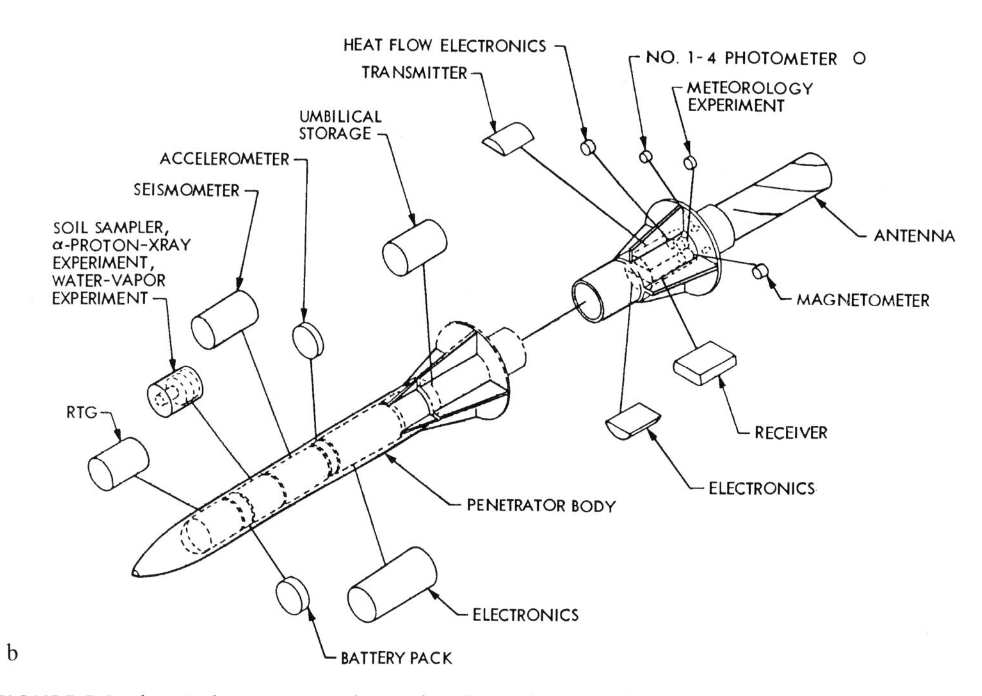

b

FIGURE 5.1: A typical penetrator to be used in the exploration of Mars will be launched from an orbiting vehicle and plunge dart-like into the Martian surface. a) For some types of instrumentations it may be necessary to employ aerobraking shrouds, as shown, to slow the descent to the surface. b) Diagram to show the various parts of a typical penetrator for planetary exploration. (NASA-Ames)

capabilities of the spacecraft and the landing and roving capabilities of the mission. This means that by the time the analysis is in progress on Mars as many as ten years may have elapsed. During those ten years there are invariably major advances made in the design of analytical instruments. These advances cannot be translated into the instruments on the Martian surface, but they can be applied in terrestrial laboratories if samples are returned from the Red Planet.

The ideal situation seems to be, therefore, to concentrate exploratory activities on first selecting sites, checking on these sites by penetrators, and then sending sample return rover missions to the most promising sites. There gather samples and return the most promising samples to Earth for immediate analysis. Some samples should be stored for future analysis when instruments are further improved.

The Soviets, by contrast, although suffering from intense political rivalry from different science groups, because of greater launch capabilities can often change payloads just before a launch, adding new updated instruments to their spacecraft. In Soviet eyes, nevertheless, this capability to place the most up-to-date instruments on their spacecraft does not invalidate the need for return of samples by a sample return mission.

The Space Science Board, following the *Viking* results, recommended that the U.S. strategy for biological exploration of Mars should consist of first an initial phase of exploration to find out which locations on Mars can provide samples which possess attributes of biological importance. This would require at least one mission, possibly two, using penetrators and a detailed survey by orbiters of the Martian surface at high resolution. Next, a decision would need to be made if these characteristics of the samples and the surface appear sufficiently promising to warrant further detailed biological exploration. A soft lander mission with rover capabilities and surface sampling and analysis may be needed for this purpose. Finally, the detailed exploration using an unsterilized sample return would investigate whether the Martian surface contains evidence of current or extinct life. In either case the obvious next activity will be a manned mission to investigate this life in great detail for comparative biology with terrestrial life and its evolution. Perhaps we might even find the common ancestor of terrestrial life forms still intact on Mars; that is, the microbial equivalent of the "missing link"; the evidence of the first living thing which connected prebiotic molecules to self-replicating molecules. As mentioned before, this evidence has been lost in the turmoil and reprocessing of Earth's crust.

During the early stages of the exploration of Mars each new piece of information returned from the unmanned spacecraft increased our understanding of the planet. The *Viking* search for life there demonstrated the inadequacy of machine intelligence. The search for life experiments could only search in certain ways. Their answer turned out to be negative insofar as the criteria decided in advance for the search.

What can be done to get the United States back into interplanetary exploration as far as Mars is concerned. A number of projects have been suggested paralleling the Russian Mars program. These logically begin with a Mars orbiting spacecraft which will circle the planet and make global maps of geological, geochemical, atmospheric, and magnetic properties.

Another great tool for further exploration is use to sensor outposts consisting of relatively inexpensive penetrators that plunge from orbit to various regions of Mars and act as remote stations to gather data about the surface and Martian weather.

The next stage of exploration is the use of a lander spacecraft carrying automated or remotely controlled rovers to the surface. Each rover will carry a variety of instruments to explore the surface around the landing site, possibly to pick up samples of the surface materials for sending back to Earth. The sample return mission would gather samples on Mars and return them to Earth as the Soviets did on the Moon and intend to do on Mars. There are several ways of accomplishing this extremely important sample return mission as are described later.

Remotely controlled aircraft, deployed from an orbiter, could be designed to carry cameras and remote sensing instruments to explore even larger regions of the surface than can be covered by rovers. They would use technology developed for high-altitude Earth surveillance aircraft and terrain following cruise missiles.

A manned voyage to Mars will be able to use new technologies developed during the 1990s, such as solar electric and solar sail propulsion systems, aerobraking and aerocapture, ability to build structures in orbit, orbital rendezvous, coupled with space operational experience gained in the *Apollo* program, with the Space Shuttle, and in the Space Station programs.

Life-supporting systems are also being developed for use in the U.S. Space Station *Freedom* (figure 5.2). Experience has also been built up on the psychological and physiological impact of long-term exposure to the space environment, principally by the Russians but to a lesser extent with varied crews of Space Shuttle missions.

There have been suggestions for hybrid missions benefiting from human and automated robots to explore Mars. Crews might be sent to Martian orbit on a routine basis like the *Apollo* program. Some would use remotely controlled vehicles on Mars' surface or in the Martian atmosphere, and some would make landings at specific sites chosen by observation from orbit. At the landing sites also there would be use of both manned exploration and remotely controlled rovers and aircraft to gather data about the area surrounding each site.

The survival of our species might ultimately depend upon mounting a serious program to establish bases and human settlements, possibly colonies, on the Red Planet. It appears that in the past comets and asteroids have collided with all the terrestrial planets. There is no assurance that these catastrophic collisions will not take place in the future. By having people on both Earth and Mars, and possibly on the Moon, there is an increased chance that humankind can survive this and other natural catastrophes.

Another intriguing thought is that Mars, who was the Roman god of war, might today offer a means to bring about peace on Earth by uniting humankind in an international effort to colonize the Red Planet. This effort would provide a common purpose for our species and generate productive and stimulating jobs for millions of people here on Earth. Rather than currently wasting national resources worldwide on military preparedness against Orwellian enemies, humankind could apply these same resources to invest its capital, the surplus of production over consumption, in a positive way that would increase the wealth of all mankind. Moreover, these new assets of an international space age would be designed for permanence rather than explosive destruction.

The development of Mars for permanent human habitation also offers a gateway to the asteroids, that enormous resource potential of small worlds revolving around the Sun between the orbits of Mars and Jupiter. These worlds offer an economic commons of enormous wealth and their mining could become the most productive human activity of the twenty-first century. Indeed, the Soviets in cooperation with France have an important asteroid mission *(Vesta)* planned for the 1990s, and the United States is also

FIGURE 5.2: Life support system that will be essential for the manned missions to Mars are being developed for use in the U.S. space station. This artwork shows how a space station laboratory module may look when deployed. Personnel will work in a "shirtsleeves" environment, and modules will be connected by tunnels as is envisioned for a Mars base also. (NASA/Boeing)

planning a comet rendezvous and asteroid mission *(CRAF)* to gather information about these bodies.

The case for human habitation on Mars is not one of waste of terrestrial resources but rather an investment of these resources. This investment could very well thrust humankind from its aggressive planetary adolescence toward its coming of age as an interplanetary species no longer confined to a limited ecosphere of a single small planet; and no longer concentrating all its high technology efforts on military projects.

The United States has for some time had the capability of sending inexpensive missions to Mars based on the *Pioneer* class spacecraft that so effectively and inexpensively explored Venus and the outer Solar System. Potential missions include an upper atmosphere and ionosphere orbiter similar to the Venus *Orbiter* and the *Atmosphere Explorer* satellites, a geochemical orbiter, a geophysical orbiter and penetrators, and an expedition to map the water resources on the planet.

A Mars upper atmosphere and ionosphere orbiter is essential to determine the structure and time behavior of the upper atmosphere and ionosphere of the planet and to determine the interactions of the solar wind with Mars and its ionosphere. Additionally, and most important, such a spacecraft would measure the intrinsic magnetic field of Mars, if any exists, and determine the shape and the flow of plasma around the planet. Photochemical processes in the upper atmosphere would be identified together with the sources and sinks of energy in the upper atmosphere. Using information about the present composition and dynamics of the atmosphere, scientists would be able to determine more accurately the evolutionary history of Mars. Also they would measure the temperature distribution in the Martian atmosphere and determine its relationship to the dynamics of dust storms, cloud formation, and changes to the polar caps. Perhaps of great importance to future exploration of Mars, this atmosphere penetrating satellite would determine the characteristics of the atmosphere and their variations. This is vital information needed to design vehicles that will have to use aerobraking techniques to enter a Martian capture orbit. Such aerobraking is critical to manned exploration of Mars with technology expected during the next quarter century.

A Mars geophysical orbiter can determine the internal structure and physical state of the planet. The average heat flow at the Martian surface would be measured, and the global figure of the planet determined more accurately. This latter is very important for landing large vehicles accurately at chosen locations on Mars. The gravity field would be measured and relationships between anomalies in the field and surface features would be identified. The global magnetic field and surface magnetic anomalies would be mapped in detail.

A geochemical orbiter and penetrators (figure 5.3) can determine the major mineral and chemical components of the surface and their global distribution. The relationship of the chemical and minerological composition to geomorphology would be determined. The concentrations and distributions of the radioactive elements, potassium, thorium, and uranium on the Martian surface would be measured. Finally, the project would measure the distribution of hydrated minerals, carbon dioxide frost, and water frost on the surface. All this information is vital to planning long-range exploration of Mars and a permanent human presence there.

Each penetrator would carry a seismometer, a heat flow thermocouple, a magnetometer, an accelerometer, a meteorology package and a soil sampler. The latter would consist of an alpha-proton-x-ray instrument and a water vapor detector. The orbiter associated with this mission would carry a radar altimeter, a gravity gradiometer, a microwave radiometer, and an X- or S-band transponder.

Perhaps the most important inexpensive *Pioneer* class mission to Mars would be the Mars water mission (figure 5.4). Its scientific objectives would be to measure the amount of water vapor in the atmosphere and the amount of condensate in the clouds. It would also determine the distribution of water frost on the surface and the distribution of hydrated minerals on the surface. It would be instrumented to search for the presence of water beneath the surface, vital information for subsequent manned missions. The mission would also measure the distribution of carbon dioxide frost on the surface, and the amount of carbon dioxide condensates in the clouds. It would ascertain the amount of atomic hydrogen escaping into space from Mars and measure the amount of ozone in the Martian atmosphere.

A study team was formed early in 1981 and several spacecraft suggested by manufacturers were analyzed. A report was issued in August 1981. Then, along with many other proposals for a revitalized American space exploration program, it was pidgeonholed in Washington.

The spacecraft for such inexpensive missions would carry a payload requiring both spinning and despun platforms. It would use rocket propulsion to decelerate from its Earth–Mars trajectory into a circular orbit around Mars. The Earth pointing communications antenna would be on the despun portion. All science data would be stored on board and periodically telemetered to Earth at 500 to 3000 bits per second.

For a Mars observer mission a circular polar orbit was chosen to allow measurements near the poles during daylight and night conditions. At a height of 186 miles (300 km) above the surface, the spacecraft would have orbited Mars in 1.9 hours. A sun-synchronous orbit was desired to simplify analysis of the scientific data and to provide a compromise between design of instruments and design of the spacecraft. The orbit for this is inclined 92.5 degrees to the equator. Local time chosen for an acceptable trade-off of science coverage and spacecraft thermal/power requirements was 9:15 A.M.

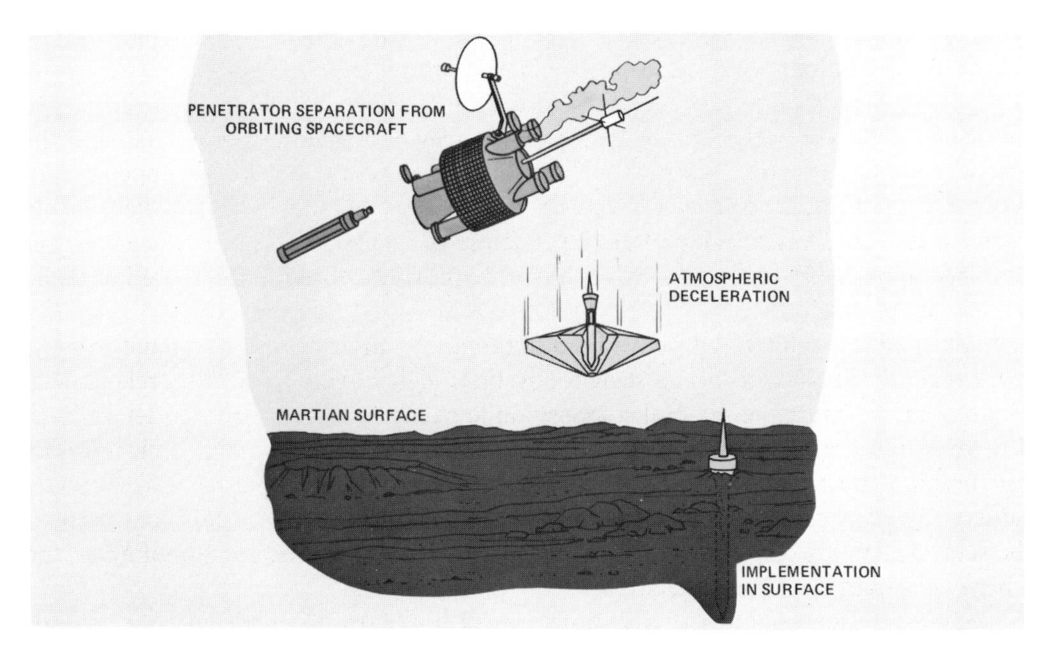

FIGURE 5.3: A geophysical orbiter spacecraft can also deploy penetrators to gain an understanding of the mineral and chemical composition of the planet's surface. (NASA-Ames)

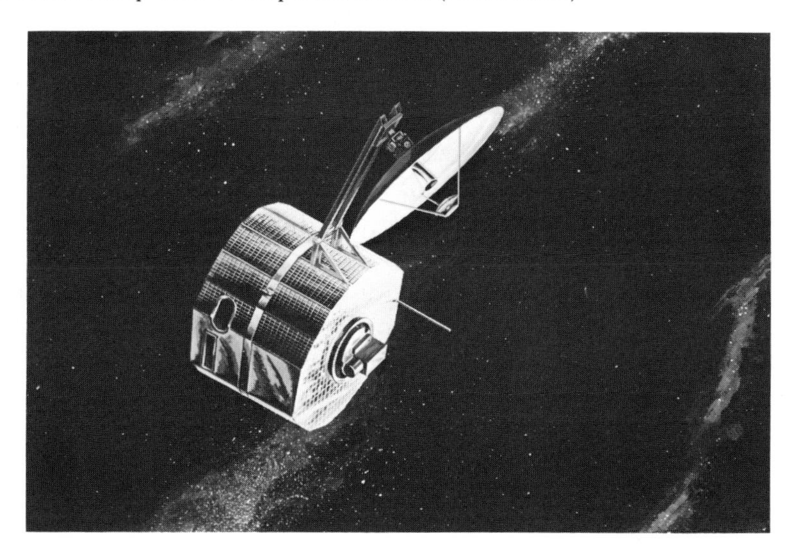

FIGURE 5.4: Another important precursor mission to Mars is an inexpensive Orbiter Water Mission to determine where water and ice are located on the planet. This mission, although not yet approved, has become increasingly important since the water detection instrument was not included on the 1992 Mars *Observer* because of cost cutting. (NASA-Ames)

The 1981 study suggested a first launch in 1988, 1990, or 1992. The proposal was to launch from the Space Shuttle using a spin-stabilized upper stage. The lifetime in orbit around Mars would be one Martian year (684 days). The spin axis would be perpendicular to the plane of the orbit, and the spin rate in orbit of the spinning part of the spacecraft would be about 5 rpm. A drawing of a spacecraft for such a mission is shown in figure 5.5. The study concluded that there were four feasible spacecraft designs based on existing hardware. The cost of the spacecraft was estimated as less

FIGURE 5.5: Bristling with sensors and cameras, the Mars *Observer* spacecraft is depicted in this drawing. The satellite is being adapted from an earlier Earth satellite. (RCA)

than $60 million, the total mission including project management and the scientific instruments and the spacecraft about $100 million.

Each of these precursor missions to Mars could have shared a design of a *Pioneer*-class spacecraft and be accomplished for under $200 million. Taking into account the time needed to implement these important missions to continue after *Viking*, the peak funding in any year would have been only $55 million and the average funding would have been less than $40 million per year for ten years. These were such insignificant amounts compared with futile expenditures on "blue-sky" military projects during the same years that it seems absurd that these missions were not undertaken. Not only did this exploration of Mars never come to pass, but the whole planetary space program of the United States languished in the doldrums. The time which elapsed between the first Soviet spacecraft to Mars and *Viking* was thirteen years. From *Viking* to the next Martian expedition (the Russian *Phobos*) was another thirteen years of inactivity. But from *Viking* to a next U.S. mission there will have lapsed the amazing period of seventeen years of inactivity. In the context of history our descendants will surely marvel that in these same seventeen years the nations of our planet spent about $8 trillion on weapons systems for "defense" of which the United States set an example of spending some $2 trillion without any real evidence that the world had been made a safer place for the human species.

One of the contractors involved in this early Mars mission study was RCA Astro-Electronics, who evaluated a polar orbiting spacecraft based on existing spacecraft design. This study contract later led to the next U.S. Martian expedition, the *Mars Polar Orbiter* or *Mars Observer* which, after many years of indecision, was finally approved for a 1992 launching and a U.S. return to the Red Planet.

Mars Observer is the first of a series of U.S. missions that will use a new class of spacecraft which has been named *Planetary Observer*. The idea is to have a spacecraft that, similar to the earlier *Pioneer* spacecraft, will allow relatively low-cost explorations of the terrestrial-type planets—Mercury, Venus, Moon, and Mars.

We have seen that the early missions to Mars provided a great deal of information; images of the surface down to a resolution of 600 feet (200 m) on most of the planet and of some areas to even greater detail. But there is little specific information on the detailed composition of the Martian surface, only broad inferences and sampling from the two *Viking* landing sites. We also lack a detailed assessment of the Martian atmosphere and the distribution of ancient materials that are undoubtedly exposed on the surface.

As mentioned in an earlier chapter, the atmosphere of Mars is 1/100 that of Earth but is substantial enough for many clouds to form. Carbon dioxide is dominant and acts as the main greenhouse gas. Also we have seen there is much dust and large weather systems. These weather systems appear to be driven by the dust absorbing heat. Climate conditions on Mars suffer from eccentricity and obliquity changes to its orbit, even more so than does Earth. It is believed that Earth's eccentricity and obliquity changes trigger terrestrial ice ages. Such changes would be expected to have an even greater effect on Mars because all the parameters are exaggerated. *Mars Observer* can be used to test theories about the causes of these ice ages.

After competitive bidding, the Jet Propulsion Laboratory, which manages the program for NASA, selected the Astro Electronics Division of RCA in Princeton, N.J., as the spacecraft contractor. The company's name has now been changed to GE Astro Space Division. In an attempt to follow the low-cost philosophy of the earlier *Pioneer* study, the spacecraft (figure 5.6) is based on the design of the RCA *Satcom-K* communications satellite, and it will use proven electronic subsystems from the *Television Infrared Observation Satellite* and the *Defense Meteorological Observation Satellite.* This approach reduces the cost of design, development and fabrication of the spacecraft. Engineering modifications and the addition of science instruments will transform an Earth orbiting spacecraft into a spacecraft suitable for use at other inner planets of the Solar System.

Originally it was intended that the *Mars Observer* should be launched by a Space Shuttle, but in view of the hold-up caused by the *Challenger* disaster, plans were changed so that the spacecraft could be launched by a Titan III expendable booster in 1992. Should something happen to prevent a launch in 1992, the mission would have to go in 1994 at the next Mars opportunity.

After an eleven-month cruise from Earth to Mars (figure 5.7), *Mars Observer* will be inserted into orbit around the Red Planet (figure 5.8). William Purdy, Project Manager at the Jet Propulsion Laboratory, explained that following a period of several months in an extended elliptical orbit the spacecraft will be placed into a low, nearly circular, polar orbit which will allow the spacecraft to cross the planet's equator at 2:00 p.m. and 2:00 a.m. each day, in what is referred to as a sun-synchronous orbit.

The spacecraft will be able to map the whole surface of Mars in just less than a Martian year of 687 days (figure 5.9). This is the nominal mission, but the spacecraft is expected to be operational beyond that period.

Primarily the *Mars Observer* mission is intended to provide an in-depth survey of the surface, the atmosphere, the climate, and the meteorology of Mars and its magnetic field. In accomplishing its mission the *Mars Observer* will determine the global elemental and mineralogical character of the surface materials and define the topography and the gravitational field globally. It is also expected to establish the nature of the magnetic field. Another aim is to determine the time and space distribution, the abundance, and the source and sinks of volatile material and of dust over the seasonal cycle of Mars.

FIGURE 5.6: The Mars *Observer* is shown in this artist's concept flying over the great Martian volcano, Olympus Mons. (JPL)

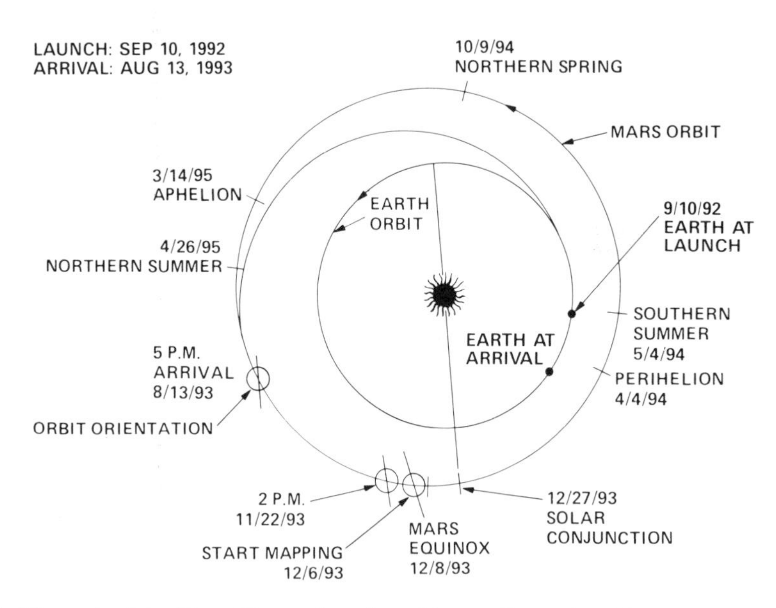

FIGURE 5.7: The path of the Mars *Observer* from launching on September 10, 1992 to arrival at Mars on August 13, 1993. (JPL)

The structure and aspects of circulation of the atmosphere will also be revealed in new detail.

There has been a suggestion that a Mars balloon radar might also be carried on the spacecraft whose radio receiver and transmitter could then act to relay the balloon data to Earth.

The characteristics of the orbit chosen are suited to all the science objectives of the mission, as explained by Frank Palluconi, deputy project scientist for the mission at the Jet Propulsion Laboratory. To determine the global characteristics of climate of Mars,

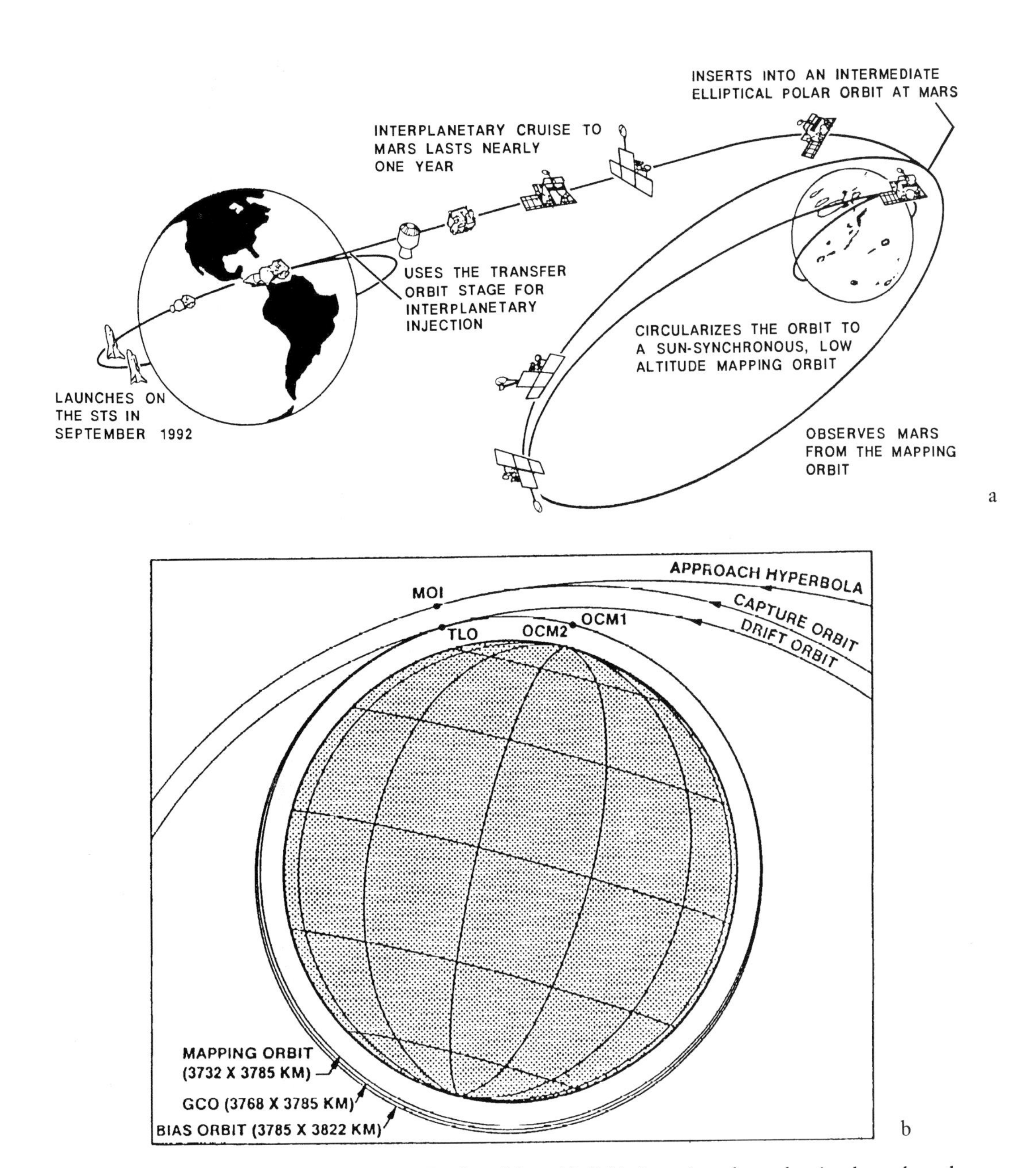

FIGURE 5.8: a) Mission highlights from Earth to Mars. b) Orbit insertion phase showing how close the spacecraft must approach Mars for it to be captured into the desired orbit. (JPL)

map the whole surface at high resolution, and achieve the other scientific objectives of the mission the spacecraft orbit has to be polar, and it has to be a circular orbit at a low altitude for high resolution of the surface. With periapsis near the south pole, he said, the orbit can be close to circular and remain stable. A sun synchronous orbit was selected to separate diurnal and seasonal effects. For this continuous surveillance of the planet, all instruments must operate all the time, so it was decided not to use a scan platform but to orient the spacecraft to the nadir. Because instruments will be gathering data continuously the spacecraft must record data continuously, so this requires data recording and playback. Data must be sent back to Earth each day at periods when the

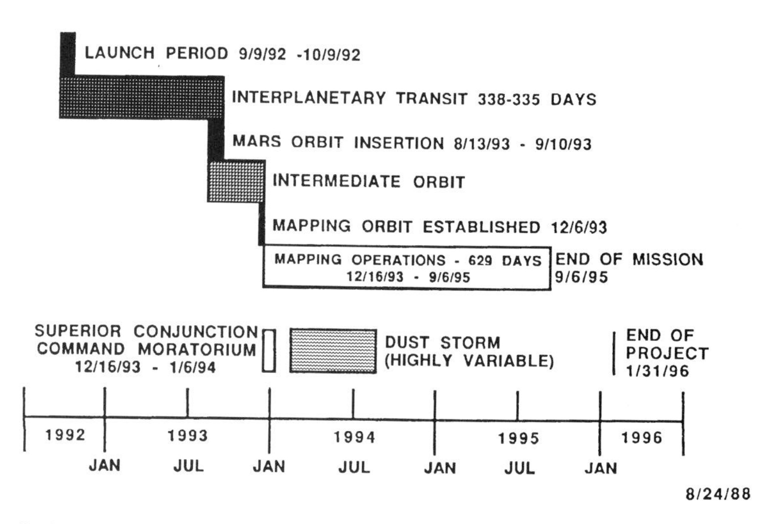

FIGURE 5.9: Outline of the Mars *Observer* mission showing the seasons of Mars and various stages of the mission profile. (JPL)

spacecraft is not behind the planet. There will also be periods of real-time data relay for some instruments. Each orbit of Mars will take 117 minutes and some data will be played back on each orbit. The spacecraft will be sending 30 billion bits of information about Mars every day for more than 600 days.

The experiments to be carried out at Mars and their principal investigators are listed in table 5.1. In addition, there are interdisciplinary scientists who will examine overlapping interests. These also are listed in the table.

A gamma ray spectrometer will measure the abundance of elements on the surface of the planet. The gamma rays from the surface of Mars originate when cosmic rays impact the surface materials. The objective of this experiment is to determine the abundance of radioactive heat source elements such as potassium, thorium, and uranium, the rock-forming elements sulfur, magnesium, calcium, aluminum, and iron, and the volatile elements such as carbon, hydrogen, and oxygen. This will provide information on which a better understanding can be gained of how the planet was internally heated in the past by such elements. This instrument is much improved over a gamma ray spectrometer used to survey the Earth's Moon. The Mars instrument is cooled to provide high spectral resolution. This instrument will also be used to detect gamma ray bursts to help find the direction of galactic sources of gamma rays.

The magnetometer carried by *Mars Observer* is important because we do not know what the magnetic field is around the planet. Earlier spacecraft that orbited Mars were too far away to get good data on the weak Martian field. The instrument consists of a magnetometer and an electron reflectometer, which measures the energy of electrons. It will probe the magnetic field at low altitude, measure the intrinsic field, determine the interaction of the solar wind with the planet, and measure local magnetic fields. Determining the magnitude and orientation of the planet's magnetic field is important to deriving other data points for validating the theory of how planets generate intrinsic fields.

The *Mars Observer* camera is a third-generation attempt to cover the annual characteristics of clouds, fogs, and frosts using a wide angle mode each day over the whole planet. Also it will provide imaging at a very high resolution (about 4.6 feet or 1.4 meters) for small parts of the planet using a line array camera which has no moving

TABLE 5.1. Experiments and Experimenters, *Mars Observer*

Instrument	*Principal Investigator*
Gamma-ray spectrometer	William V. Boynton University of Arizona
Thermal emission spectrometer	Philip R. Christensen Arizona State University
Television camera	Michael C. Malin Arizona State University
IR radiometer	Daniel J. McCleese Jet Propulsion Laboratory
Radio science	G. Leonard Taylor Stanford University
Magnetometer	Mario H. Acuna NASA Goddard Space Flight Center
Scientific Field	*Interdisciplinary Scientists*
Geoscience	Michael H. Carr U.S. Geological Survey
Surface-atmosphere interactions	Bruce M. Jakowsky University of Colorado
Atmosphere & Climatology	James B. Pollack NASA Ames Research Center
Polar atmospheric sciences	Andrew P. Ingersoll Calif. Institute of Technology
Data Management & Archives	Raymond A. Arvidson Washington University

parts. High resolution pictures of the surface are important in connection with interpreting the geological features and for planning roving missions on the surface.

A pressure-modulated infrared radiometer will make atmospheric sounding observations and limb observations. It will also be used to compile a radiation budget for the surface. Originally there was to have been a 6.2-mile (10-km) resolution radar altimeter but this was replaced by a laser altimeter with 540 feet (165 m) resolution. The instrument will provide excellent altitude profiles for contour mapping, but these will not completely cover the Martian surface.

There will be good radio science with the *Mars Observer.* Its 12 to 13 orbits per day will produce many occultations whereby small-scale atmospheric data can be obtained. The low orbit will allow a good survey of the gravity field by providing data on how the orbit is perturbed. In fact, the scientists expect *Mars Observer* to provide better gravity data for Mars than we have for the Earth.

A thermal emission spectrometer will provide a continuous spectrum from 6.25 to 50 micrometers. The spectrum of thermal infrared radiation is rich in absorption features caused by minerals in rocks. This instrument will complement the elemental abundance experiment. It will also be used to gather data about atmospheric dust and clouds. The information it will provide about the thermal physical properties of the surface will be much better than that obtained by the *Viking* orbiters and is important data needed to operate a manned base on Mars.

A visual and near infrared spectrometer originally intended for the spacecraft to furnish a mineralogical map of Mars and chart concentrations of water and carbon dioxide in the atmosphere and on the surface had to be removed from the payload; a sacrifice to cost reduction because of the two-year slippage in the launch date arising from non-availability of launch vehicles.

A U.S. mission to Mars using an automatic and remotely controlled roving vehicle to traverse the planet's surface, gather samples of that surface and return them to Earth has been recommended by the Solar System Exploration Committee of the NASA Advisory Council. This mission has the support of many scientists as a way to find out more not only about Mars but also about the origin and early evolution of our own Earth. The committee recommended that the mission should take place before the year 2000.

There have been many studies of missions to gather samples from Mars and to explore its surface with roving vehicles. An unmanned Mars Rover and Sample Return mission could be mounted with two launches of Centaur/Titan boosters in 1998. The objectives of this advanced mission proposed for the United States are to make in-situ measurements on the Martian surface and return samples of the regolith to Earth. The aim of this program is to reconstruct the geological, climatological, and biological history of Mars, to obtain key environmental information about the planet, and to test the key technologies necessary for later human exploration of Mars. The Mars Rover and Sample Return mission is a necessary precursor to any human exploration of the Red Planet. The precursor role is to obtain information on surface materials, their physical and chemical properties, the nature of the dust and if it is toxic, the atmosphere, and the surface environment. It will also find out if there is any biology on Mars before humans attempt to land on the surface and contaminate it biologically.

In 1987 a joint study by the Jet Propulsion Laboratory, the Johnson Space Center and Science Applications International Corporation examined such a U.S. mission for the mid 1990s, as a step toward U.S.-manned missions in the next century. The overall concept is to fly a vehicle to Mars that will be placed in orbit around the planet. A lander will separate and carry a rover down to the surface (figure 5.10). Directed by remote control, the rover will explore around the landing site and pick up a few pounds of samples of interest to geologists and exobiologists monitoring the mission. After about a year of exploring the surface the rover will return to the lander and transfer the samples to a small ascent rocket that can carry the samples back to the orbiter. The sample-container would be transferred in orbit to an Earth-return vehicle attached to the orbiter. This spacecraft would then separate from the orbiter and use rockets to escape from Mars orbit and start its ten-month return trip to Earth. Approaching Earth at 27,000 mph (13 km/s) the return vehicle will use rocket braking or aerocapture to attain an orbit around Earth. It could then either rendezvous with the U.S. Space Station for quarantine inspection and analysis of samples or with a Space Shuttle for direct final return of the samples to terrestrial laboratories.

Estimates of the magnitude of this mission suggest an initial mass in low Earth orbit of 9.7 tons (about 8800 kg), requiring assembly in orbit from units brought up from the Earth by two Space Shuttle flights. The interplanetary vehicle, which will become the Mars orbiter and the return flight vehicle, will weigh about 3.3 tons (3000 kg), and the lander, rover, and ascent vehicle will weigh about 6.4 tons (5800 kg). The rover will amount to 880 lb (400 kg). Of the initial weight of 9.7 tons, 5.64 tons (5120 kg) consists of propellants. The amount of samples returned to Earth in a 33 lb (15 kg) cannister will be 11 lb (5 kg).

The current phase (mid-1989) of the study of Mars Rover and Sample Return

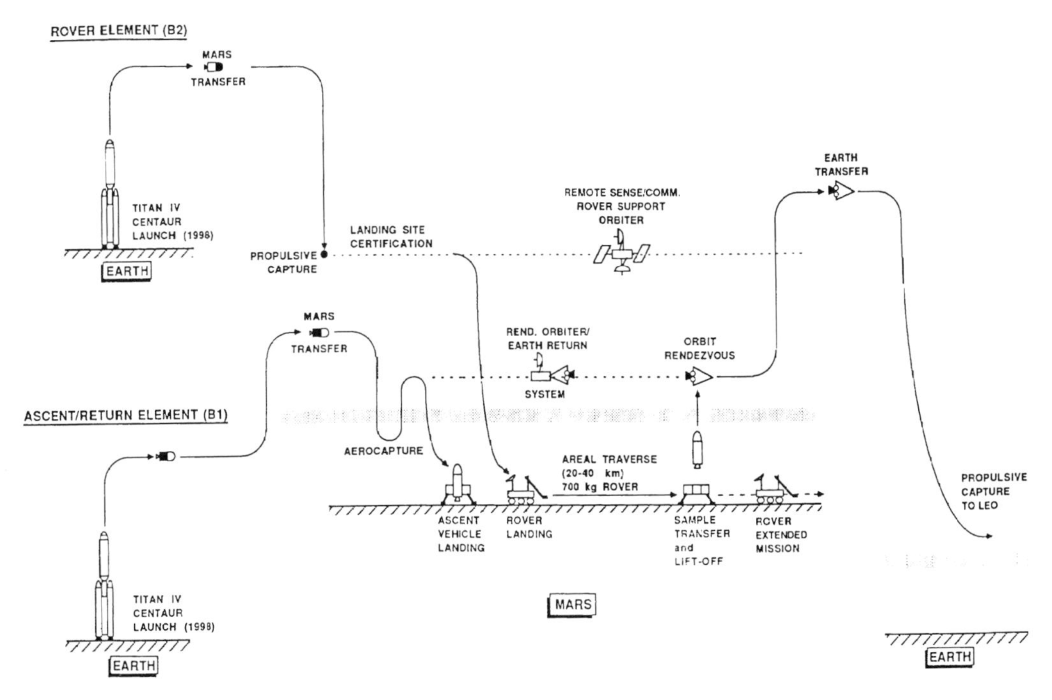

FIGURE 5.10: A typical mission profile for a sample return mission to Mars in which two Titan IV/Centaur launch vehicles send two spacecraft to Mars at a single launch window. One spacecraft has an orbiter and a lander, the other carries an ascent vehicle and an Earth return vehicle. The rover has to return to the ascent vehicle to transfer samples, but can then continue afterward in an extended exploratory mission. (JPL)

missions is termed a pre-phase A period, during which several options are being defined and evaluated. The most promising of these options will then be selected for a more detailed study.

Three launch configurations were included in the study: a heavy lift launch vehicle to launch the whole mission by one vehicle, a split launch in which the orbiter and rover are launched in one Titan-IV/Centaur-G launch vehicle, and the Mars ascent vehicle and Earth return vehicle in another. A typical arrangement of components for this configuration is shown in block form in figure 5.11. A third possible configuration also uses two Titan/Centaur launch vehicles, one containing the orbiter and Mars ascent vehicle and the other carrying the rover and the Earth return vehicle. Launching from Space Shuttles is no longer a popular option.

The Mars communication orbiter and the sample return orbiter remain in orbit around Mars while two landers carry their payload to the surface. One lander carries the rover, the other carries the Mars ascent vehicle which brings the Martian samples back to the Earth return vehicle where they are stowed in the sample return capsule for the interplanetary voyage back to Earth.

There have been studies of the use of the Martian atmosphere to slow the approaching spacecraft into a capture orbit around Mars. This aerocapture technique will save a large amount of propellant and is really essential for a meaningful mission to gather samples from the Red Planet. The spacecraft will approach Mars at a velocity of about 10,000 mph (4.46 km/sec). Without reducing this speed the spacecraft would fly by Mars in a hyperbolic path and continue past the planet into space. However, if the trajectory passes through the high atmosphere of Mars, 70 to 100 miles (110–160 km)

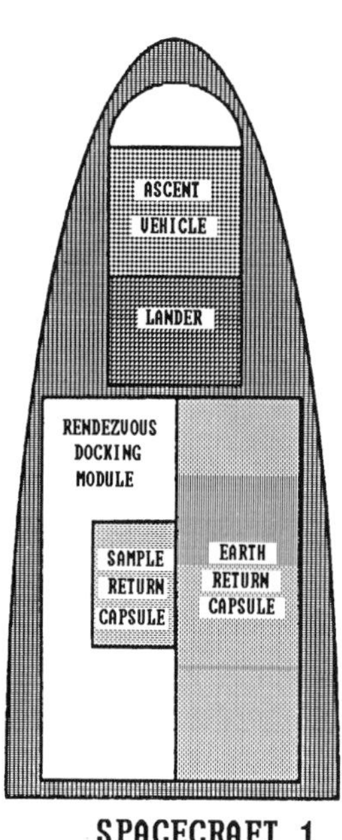

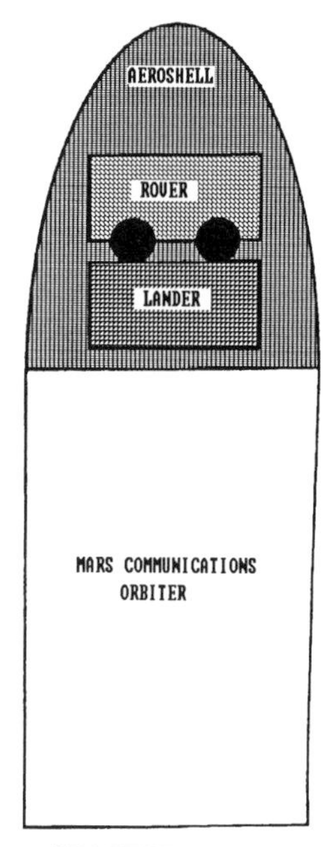

FIGURE 5.11: Block diagram of the arrangements of components for the Mars Rover and Sample Return mission using two spacecraft to reach and orbit Mars. One carries the rover and its lander, the other carries the ascent vehicle and its lander. The Earth return vehicle and sample capsule is carried in Spacecraft 1. (After NASA-Johnson study)

above the surface, atmospheric drag can be used to slow the spacecraft enough for it to be captured into an elliptical orbit. This requires an advanced guidance system for the interplanetary spacecraft for precisely controlled entry into the Martian atmosphere at the correct angle, otherwise the spacecraft could burn up or richochet off the top of the atmosphere without slowing down.

Aerocapture of the spacecraft into orbit around Mars thus requires precise navigation during approach to the planet. An onboard navigation system appears to be essential for this purpose. The line of sight to the Martian satellites can be measured against the background of stars, but because these stars will be very faint, this might be difficult to accomplish automatically. Another approach is to provide a reference frame on board the spacecraft in the form of an inertial measurement unit, and to use this to reference the positions of the satellites. This unit would be kept aligned by a star tracker sighting on bright stars.

Approach velocity to Mars will be between 2.12 and 2.77 miles per second (3.41 and 4.46 km/sec.). The navigation system needs to define the entry into the Martian atmosphere at 78 miles (125 km) altitude with an entry angle within a spread of only one degree.

There is no doubt that aerocapture into an orbit around Mars is a complex and multifaceted problem which must be solved not only for the sample return mission but also for future manned missions. One of the big uncertainties, which also applies to aerocapture at Earth, is the nature and effects of unpredictable variations in atmospheric density at high altitudes where aerobraking has to take place. An unexpected drop in density as the spacecraft skips back from the atmosphere could lead to unac-

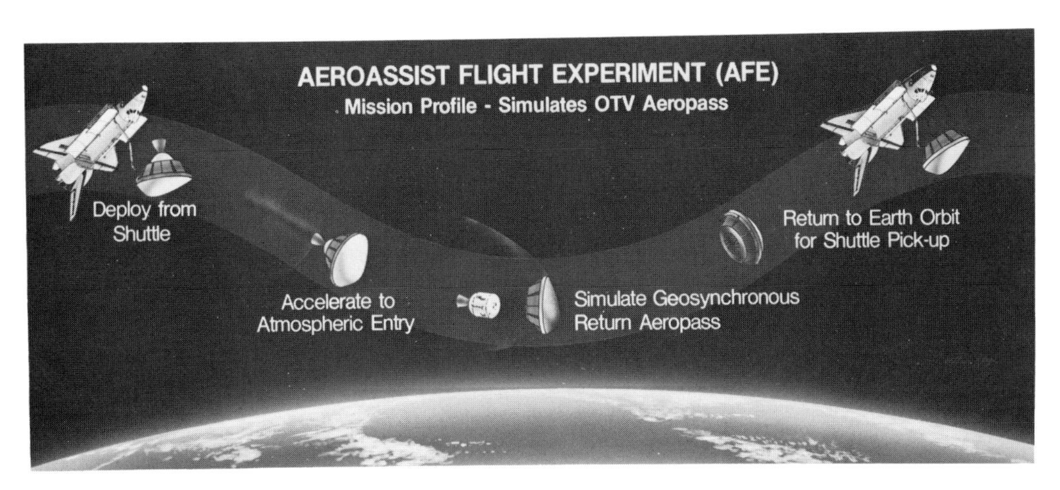

FIGURE 5.12: The Aeroassist Flight Experiment will gain scientific data for aerobraking technology. The AFE is launched from and recovered by a space shuttle. Its mission is to accelerate into the Earth's atmosphere and use aerobraking at high altitude to slow down again to orbital velocity as it gathers important information about nonequilibrium flow and aerodynamic heating at high velocity and high altitudes. (NASA-Ames)

ceptable changes to the resultant altitude of periapsis of the capture orbit. Information about how to correct for these circumstances may be forthcoming from the Aeroassist Flight Experiment (figure 5.12) planned by NASA for accelerating a test vehicle launched from the Space Shuttle to enter into the terrestrial atmosphere in 1994 and skip back to the Space Shuttle for recovery. Among its science experiments this program will check for the effects of aerodynamic forces on the control system during the aerobraking flight through Earth's atmosphere which is also known to have unpredictable variations in density at high altitudes.

Each lander has to decelerate its payload of rover and ascent vehicle during the terminal phase of descent and safely land at a desired location. Descent of the landers through the atmosphere after the orbiter has been established safely in its orbit, presents another great technical challenge. The experience with *Viking* landers is important here. The landing can be accomplished by parachute, retropropulsion, or a rotor system, or various combinations. The *Viking* landers used parachutes followed by rocket braking. So far, studies eliminate the rotor system and favor a parachute system followed by a rocket system.

After separation from each orbiter, each lander will use rockets to slow its orbital velocity and cause it to enter the atmosphere. Again precise orientation and positioning is essential to a success of the mission. An aeroshell will be used to provide some aerodynamic control of the path to a chosen landing site. Later, when the velocity has been sufficiently reduced, the aeroshell will be jettisoned and a parachute deployed at an altitude of about 16,000 feet (5,000 m). This will slow the lander until it is moving slowly enough for retrorockets to be used at about 3,300 to 6,600 feet (1,000–2,000 m) for the final descent to and soft landing on the surface.

The landing phase of the Mars Rover and Sample Return mission must ensure that the rover and an ascent vehicle (see figure 5.13) are delivered safely to a preselected location on Mars. Since television pictures of hazards to the landing will take 20 minutes or so to reach Earth, the landing sequence cannot be controlled from Earth.

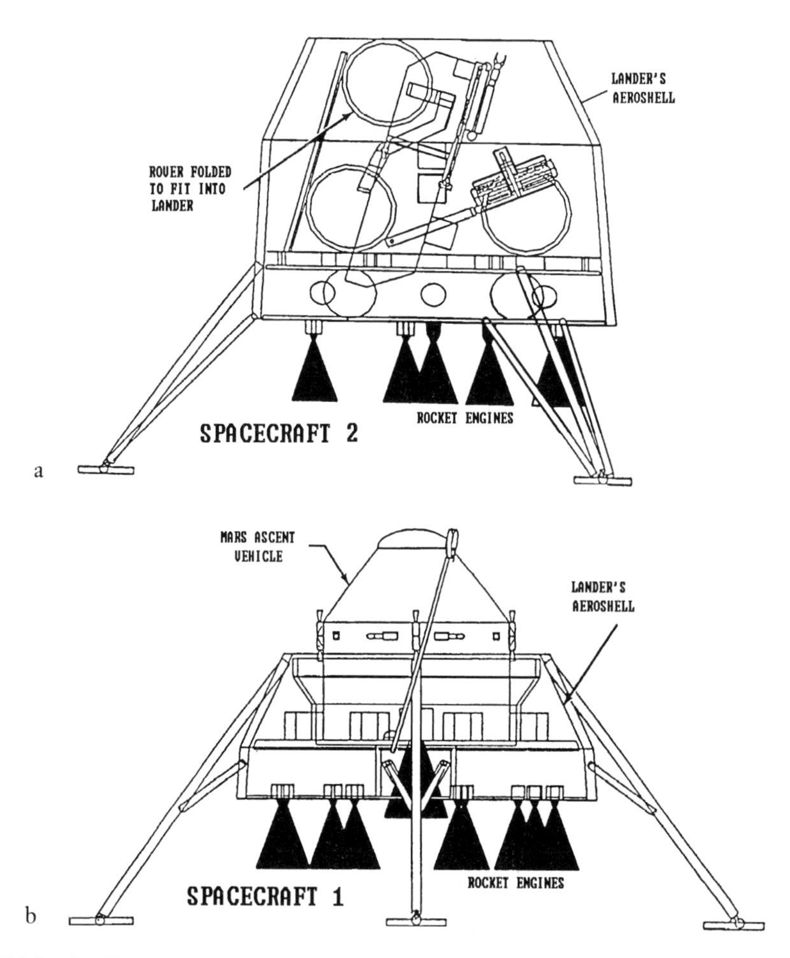

FIGURE 5.13: a): The rover is packaged within the lander as shown in this simplified drawing of a possible configuration. After the lander touches down a landing ramp is deployed and the rover moves along it onto the Martian surface. (b) This diagram shows the other lander with its ascent vehicle on the Martian surface. When the rover has gathered its samples, they are transferred to the ascent vehicle which then carries them to orbit to rendezvous with and be transferred to the sample capsule in the Earth return vehicle. (After NASA-Johnson study)

The lander spacecraft has to have its own "brains" and hazard avoidance system to guide it automatically to a safe place for a landing that will allow the rover to be deployed. For example, if the lander should go down in a field of large jagged boulders it would not only be dangerous to the lander but also could prevent the lander from unloading its roving vehicle. Delivering these payloads to a location of scientific interest while avoiding surface hazards requires use of a navigation system that can operate autonomous as did the *Viking* landing system. Autonomous operation is required because of the long-trip communication time between Mars and Earth and the possibility that communications will be interrupted during the entry phase, both direct to Earth and via an orbital relay.

An in-depth study by several NASA groups has shown that navigation errors can be significantly reduced by the lander tracking on the orbiter as was done with the *Apollo* Lunar Module tracking the Command Module for landing on the Moon. Also beacon tracking could be achieved by placing radio beacons on the surface ahead of the landing. The final position of the landing could be within 1500 feet (500 m) of a selected site by

FIGURE 5.14: One design for a Mars rover is shown in this artist's concept by Ken Hodges. The vehicle consists of three connected compartments, each mounted on a pair of wheels. It is able to negotiate obstacles such as rocks, as illustrated. It will be controlled from Earth via the orbiting spacecraft shown in the distant sky. (JPL)

using these methods. Much greater accuracy can be achieved, possibly to within a few meters, by landmark tracking on an object near the chosen landing site. But autonomous landing tracking equivalent to the *Apollo* crew's manual landmark tracking will require significant development and testing.

The lander has to be able to land in a region where surface hazards are less than 3 feet (1 meter) in diameter in order that the landing legs can be small enough to be packaged within the aeroshell needed for atmospheric braking on entry into the Martian atmosphere. A hazard recognition and avoidance system will have to be carried by the lander to meet this requirement.

The orbiter is expected to support the operations on the surface by providing imaging for a survey of the landing site and afterward by providing detailed imaging of terrain around the rover so that its path can be planned. The orbiter will also relay information from the rover and lander to Earth. While the rover might be able to send information directly to Earth at low bit rates, the orbiter provides an important backup should direct communication with the rover be interrupted. Orbiter science also has the objectives of determining planetwide surface characteristics, evolution sequence, volatile history, interaction of the surface with the atmosphere, structure and dynamics of the interior, and to search for prebiotic and biotic fossils of extinct life forms.

The rover is expected to have a mass of up to 3300 lb (1500 kg) and be able to travel a distance of a few miles each day. Several different configurations have been suggested. One is for the rover to consist of three two-wheeled cabs flexibly connected so that each cab can move in three axes of pitch, yaw, and roll (figure 5.14). Each wheel will be about 3 feet in diameter. This version of a rover could climb over 5-feet high obstacles, and climb grades of up to 30 degrees, depending on the compactness of the soil. One cab would contain instruments to obtain information about the surface and would have a drill mechanism and robot arms to scoop up samples. The middle cab would contain communications and control and navigation systems, and the third cab would provide power for the rover from a radioisotope thermoelectric generator. This type of roving vehicle would be steered by rotating the two end cabs relative to the center cab. A stereo vision system mounted on a mast would be used to search out the desired path and navigate the rover along it.

The U.S. Army has been experimenting with teleoperated robot vehicles. In 1988 Grumman and Martin Marietta received contracts to develop and field test a Tactical Multipurpose Automated Platform (TMAP), for military use. This program will help to develop technology that could be applied to Martian rovers. Earlier, Grumman built and tested with the Army a 380-lb (172-kg) vehicle in 1984 which could travel off roads at 10 miles per hour (16 km/hr) to fulfill surveillance missions with a zoom optical system and demolition missions using guns and missiles. The more modern version which is expected to weigh about 600 lb (272 kg) is designed to negotiate 12-inch (30 cm) steps or trenches, climb a 31-degree slope, and turn in place. A simple articulated chassis design keeps all the four wheels level on uneven surfaces. Soldiers have already learned how to operate such vehicles from several thousand yards away. This military experience will, of course, be useful in developing methods to control and operate rovers on another planet.

Martin Marietta Space Systems has been actively involved in various aspects of missions to Mars following the company's major effort in the *Viking* mission. Under a contract with the Jet Propulsion Laboratory, Martin Marietta is studying a concept called the "walking beam" for NASA's Mars Rover and Sample Return mission planned for 1998. The walking beam (figure 5.15) is one of several wheeled and walking designs being considered by the company. The vehicle is a collapsible, seven-legged T-beam structure that will be stored in the lander spacecraft. It will travel across the Martian surface at about 300 feet per hour using its long telescoping legs to step up to five feet to climb obstacles.

For the rover mission also Martin Marietta has created a scale model of the Martian surface (figure 5.16) to help engineers determine whether an unmanned spacecraft can choose a landing site on its own. A television camera mounted on a moving robotic arm simulates a spacecraft approaching the surface. A high-speed computer interprets the television images, telling the spacecraft how to avoid hazards such as boulders and craters as it descends to a landing.

Several methods of controlling the Mars rover have been suggested. A computer-aided remote driving method relies on images from the camera system to designate a path of some 800 feet (250 m) in length, correlated with analysis of images from the orbiter. Since some obstacles might not be visible on the images from orbit, a radar system on the rover would automatically stop movement should it detect an unforeseen obstacle. Another technique of the navigation and exploration is semi-autonomous. High resolution images from the orbiter will be relied upon to issue a command to the rover for a longer traverse each day. The rover would complete the traverse and then wait for another command. Again it could use a radar fail-safe system to prevent collision with unforeseen hazards. The rover might also be able to use technology developed for cruise missiles.

Three kinds of traverse sequences have been considered: local to less than 330 feet (100 m) from the lander, areal to cover up to 25 miles (40 km) from the lander, and regional to cover more than 60 miles (100 km) from the lander. The areal is popular because it fits better within the payload expected to be available and the technology anticipated.

The rover (figure 5.17) needs to be able to drill into the surface to obtain fresh rocks as well as rocks and pebbles collected from the surface. An important discovery would

FIGURE 5.15: Artist's concept of a "Walking Beam" Mars rover being designed and evaluated by Martin Marietta as an alternative to wheeled vehicles for roving the Martian surface. (Martin Marietta)

FIGURE 5.16: A scale model of the Martian surface is being used by engineers at Martin-Marietta, Denver, to determine how unmanned spacecraft can be instrumented to avoid obstacles when landing automatically on Mars. (Martin Marietta)

be to find subsurface ice, and this is only possible through drilling. The aim in gathering samples is to obtain as much variety as possible; mass is not that important because there are nowadays very sophisticated analysis instruments that require only very small samples.

In 1985 a mission study was performed by the Jet Propulsion Laboratory based on orbital rendezvous as used by *Apollo*. In this case study two spacecraft traveled from

FIGURE 5.17: This artist's concept illustrates an automated and remote-controlled rover sampling the Martian surface with the lander vehicle in the right background. The rover would include a stereo camera vision system, sensors, a computer "brain," controlled manipulators, and a drill system for acquiring samples. This concept was about 18 feet long and 6 feet across, weighing some 1540 pounds (700 kg). (NASA-Johnson)

Earth to Mars. A lander vehicle would have been carried to Mars on a carrier orbiter spacecraft. The lander would have separated and automatically descended to the planet's surface where it would have picked up samples of Martian soil and stored them in a return capsule. The lander did not carry a rover so samples could only be obtained close to the lander. This mission would have suffered from the same limitations as the *Viking* sampling technique, but the aim was to return samples to Earth where they could be more thoroughly examined and analyzed than could be accomplished with an automated laboratory on the Martian surface similar to the type carried by the two *Vikings*. A return spacecraft would stay in a 300-mile (500-km) orbit around Mars. The lander consisted of two parts, a base structure and an ascent vehicle. The base structure serves as a launch pad for the ascent vehicle. This structure would have been left behind (figure 5.18) as was the case with part of the Lunar Excursion Module of *Apollo*. After completing the surface mission the ascent stage of the lander would reach orbit and rendezvous with the return vehicle orbiting Mars. Samples would be transferred in a capsule from the ascent vehicle to the return vehicle which would carry the samples back to Earth. This initial study has now been expanded to include a rover so that a more varied sampling can be obtained.

Aerocapture of the spacecraft carrying the samples on its return to Earth is important if the initial mass in Low Earth Orbit (LEO) required for the Mars Rover and Sample Return mission is to be kept at a reasonable amount. The heating rates on aerocapture through Earth's upper atmosphere into LEO are such that the thermal protection system must be an ablator as it would have to be for a direct entry into the atmosphere followed by a landing. The deceleration forces are, however, much higher for a direct entry than for an aerobraking return to LEO. If the sample return capsule cannot withstand forces in excess of five times that of gravity, then aerocapture will have to be used. For this purpose of aerocapture or direct entry a biconic high lift/low drag vehicle does not appear to have significant performance gains over a blunt-nosed low lift/high drag vehicle similar to the *Apollo* capsule (figure 5.19).

There are, however, other pros and cons for these two classes of entry vehicles. The low lift/drag vehicles have minimum capability for aerodynamic maneuvering but can

FIGURE 5.18: An early proposal for obtaining samples from Mars did not include a rover. A lander, based on *Apollo* technology, descended from an orbiter and gathered samples at the landing site. These were to be transferred to the ascent part of the lander which then carried them to an Earth return vehicle which had remained in orbit around Mars during the lander's mission. (NASA-JPL)

carry more payload for a given volume. The high lift/drag vehicles have better maneuverability but pack less payload in the same volume. One suggestion from recent studies is a combination vehicle which is shaped as a symmetrical biconic with a lift/drag ratio between the blunt-nose and the bent biconic vehicle.

NASA's Johnson Space Center designed a direct entry sample return capsule using a scaled-down version of the *Apollo* capsule. The thermal protection system for a direct entry capsule has to be about 16 percent greater mass than for an aerocapture because the direct entry leads to additional heating. After entry, the return capsule, similar to the *Apollo* capsule would descend by parachute and be designed to be air snatched into a pickup airplane as were some surveillance satellites in early military programs. Air snatch avoids ocean splash down and possible loss. Typical cross sections of an aeroassist return capsule and a direct entry capsule are shown in figure 5.20.

Roger Bourke, study manager of the Mars Rover and Sample Return missions, at the Jet Propulsion Laboratory, said in a recent interview that these missions are very complex undertakings and push to the limits available in terms of mass landed on Mars and mass returned to Earth, time on Mars, complexity of hardware, technology, and cost. The demands can be reduced by stretching time which is achieved by spreading the mission over several launch windows (figure 5.21); that is, launch separate parts of the mission in 1996, 1998, and 2001. The concept gaining favor is to use a mapping orbiter launched in 1996 to select and certify the site for the landing, taking two years to do this if necessary. In 1998 send the lander and the rover and allow these to operate for two years on the surface. Finally, in 2001 send the ascent and return vehicle. The ascent vehicle would be placed down near to the then current position of the rover.

a

b

FIGURE 5.19: Two types of entry vehicles for aerobraking in an atmosphere. a) A high lift/low drag vehicle provides greatest maneuverability but can carry less payload. b) A blunt-nosed low lift/high drag vehicle carries a higher payload at the expense of maneuverability. (NASA-Ames)

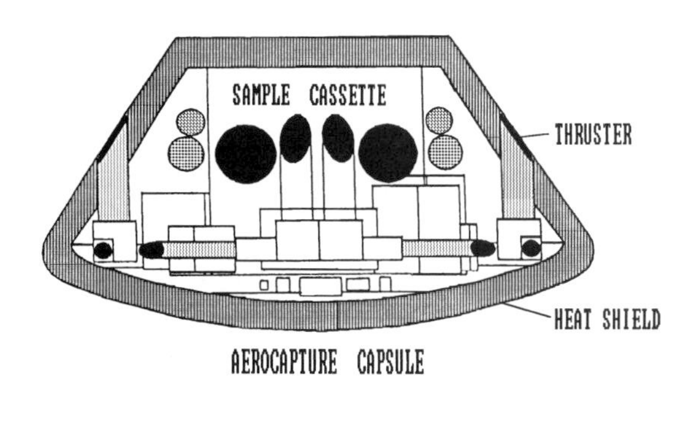

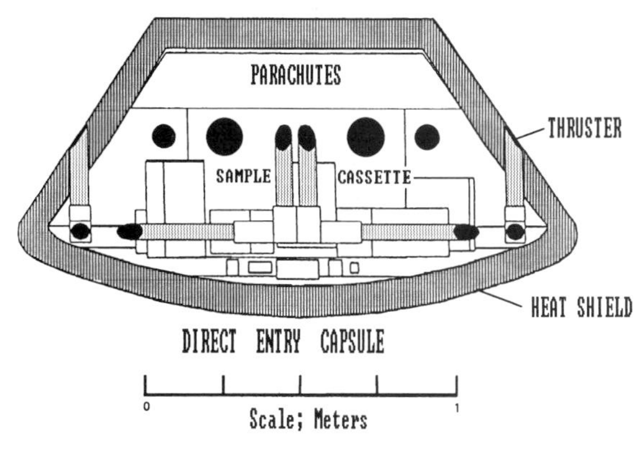

FIGURE 5.20: Cross section of an aeroassist and a direct entry sample return capsule for bringing Martian samples safely back to Earth. (After NASA-Johnson study)

This would accordingly allow the rover to travel much further from the initial landing site than it could if it had to return to the initial landing site to send its samples back to Earth. With this mission the rover might be able to travel some 500 miles (800 km) from its landed site.

Why do we need samples of the surface? Many measurements cannot be done remotely, for example, age dating of rocks, the composition of stable isotopes, and the amounts of trace elements. Making the experiments in situ on Mars in the U.S. space program as currently operating requires that the measurements have to be decided on about ten years in advance of their being made. By contrast, a return of samples, as with the Moon rocks, allows scientists to take advantage of later improved techniques to analyze the rocks. Also, the samples can be made available to a much wider scientific community for tests. Moreover, as with the Moon samples, Mars samples can be placed in storage for future analyses when analytical methods are further improved.

Why do we need a rover? A variety of samples is required from a number of sites. A rover aids the selection of samples. It also permits in situ measurements of environmentally sensitive samples. We need also to document, as we did on the Moon, the geologic context of the area from which samples are collected; such things as lava flows, rock glaciers, river valleys, and the like. Also, observations at the scale permitted by a lander will provide insights into landscape forming processes.

Many science questions need answers. Several refer to formation of planets. What were the temperature, pressure, and chemical variations prevailing in the Solar System when the planets were formed? When and how did Mars accrete its materials compared with the other planets and with meteorites? Was accretion homogeneous or inhomogeneous? What was the thermal state of Mars at the end of accretion? Did internal

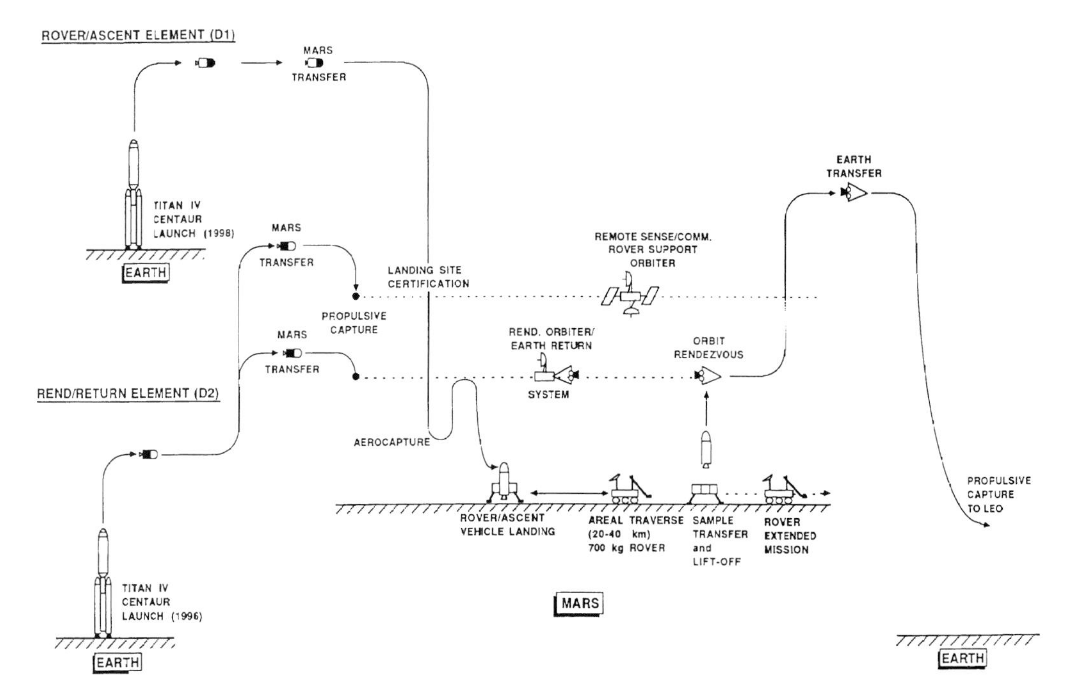

FIGURE 5.21: A favored sample return mission profile is depicted in this drawing. One spacecraft contains the three modules of lander, rover, and ascent vehicle, the other contains two modules. These latter are a combination of orbiter and support spacecraft, and an orbit rendezvous and Earth return spacecraft needed to bring samples back to Earth. The orbiter/return spacecraft is launched first and the two spacecraft are placed in orbit around Mars from which the landing site is selected. When this has been done, hopefully before the next launch window—i.e., within two years—the second spacecraft is dispatched to land an ascent vehicle and the rover on Mars. When the sample mission has been completed, the rover transfers the samples to the ascent vehicle which carries them into orbit where they are transferred to the Earth return vehicle and taken back to the Earth. As with the other launch profile, the rover can continue on an extended mission. (JPL)

fractionation occur while the planet was still accreting? Concerning the internal structure, we would like to know the comparative dimensions of the crust, mantle, and core of Mars, and when the global fractures occurred.

Questions about geologic evolution include the timing and nature of volcanic activity. We need to confirm the nature and origin of near surface materials. What is the history of the impacts that gouged out the many sizes of craters on Mars? What are the erosional, depositional, and weathering processes on the planet today, and what might they have been in the past? What is the history of volatiles on the planet? There is currently a wide range of opinions on the amount of volatiles outgassed from the interior of Mars over its history. We need to resolve the differences if we are to understand planetary evolution. Do hydrogen, helium, and noble gases indicate a low volatile inventory on Mars despite the fact that erosional episodes suggest as much water as on Earth, relative to carbon dioxide and nitrogen?

Another important question is when did the polar deposits accumulate?

Going to the right places in the ancient cratered terrain, for example, landing in crater lake sediments, should help to reveal the history of sedimentary deposits. Another important question is whether or not there are major sinks of water on Mars and if so, where? Also are there carbonates and nitrates?'

As for the atmosphere; we want to determine the climatic history of the planet. Does the atmosphere change with time? Was liquid water stable on the surface? When? Where? What controls the dynamics of the atmosphere? Has this changed with time?

How did Mars lose most of its atmosphere is another important question needing to be resolved. Was the loss because of greater impact rates arising from the planet's proximity to the asteroid belt? The evidence is ambiguous. One recent paper suggests that Mars was a wet planet with a thick atmosphere early in its history. Water scavenged the carbon dioxide from the atmosphere and volcanic activity pumped the carbon dioxide back into the atmosphere. Other planetologists are skeptical of this theory, based on the fact that there is no evidence of a rapid decrease of volcanic activity that would allow the carbon dioxide to be stored in the Martian surface. However, the Martian volcanoes may be still active today; there is certainly evidence of fresh lava flows with no impact craters on their surfaces. The channels on Mars appear to have been carved early in the history of the planet, thereby indicating that large amounts of liquid water did not last long after the planet's formation. Carbon dioxide could be returned into the atmosphere during the period of impacts because these would dissociate carbon dioxide from the rocks. When impacts tailed off the atmosphere would be recaptured into the ground.

On the history of obliquity: to resolve this we need a rover mission to the polar regions. Technically this is extremely difficult. How would such a vehicle operate over 100 miles on frost-covered ground? Orbital support and delivery of the rover to polar regions would be extremely difficult. So far, studies have concentrated on equatorial missions as the most technically feasible. A big technical challenge is automating the rover, and how to sequence its exploration of the surface.

Site selection is important, and a number of candidate sites were suggested in 1980 by a Mars Science Working Group Site Selection Team. Layered rocks which might be sediments could be obtained from Candor or Hebes Chasma. Old crustal rocks might be obtained from landings on Tyrrhena Terra or from Iapygia. Sediments from fluvial channels can be collected in Chryse Planitia. Young volcanic rocks can be sampled on Arsia Mons or Apollinaris Patera. Finally, polar ice and deposits could be sampled by landing within 6 degrees from the north pole.

What about biological contamination of Earth from returned samples? It is possible to do all the biological testing for hazardous life forms at the Space Station, but this only moves the problem slightly away form Earth. The question would be what to do if the samples did contaminate the Space Station. Abandon it? A return directly to Earth laboratories is preferred because greater control is possible on Earth than in the Space Station.

Efforts to prevent contamination of planets by spacecraft carrying terrestrial life forms to them began in 1956 when a meeting of the International Astronautical Federation drafted some suggested rules. Over the years more definitive policies have been developed not only to prevent other planets from being contaminated by terrestrial life forms but also to prevent any back contamination of the terrestrial environment by alien life forms from other worlds. This latter will assume greater importance when the Mars sample return missions begin. However, the scientific consensus is that the problem is more of a social and political issue rather than a scientific issue in view of what has already been discovered by *Viking* about the Martian surface.

Nevertheless scientists are especially concerned about preserving the pristine nature of samples so that their scientific value will not be reduced. Preserving the samples for scientific analysis will call for more extensive controls than preventing them from affecting the terrestrial environment. In May 1988, the latest planetary protection policy established anew that controls on contamination always be imposed, and re-emphasized NASA's commitment to preserving planetary conditions for future exploration.

The *Viking* landers were sterilized so as not to contaminate Mars; future landers can be treated likewise. However, the return of samples to Earth will present new problems needing solution. The sample return canister must be sealed effectively and it must not carry on its outer surface any material from Mars. Particularly troublesome will be eliminating Martian dust which might prevent effective closing of seals. All dust will have to be removed from the canister and its seals. None must be carried to the Earth transfer return vehicle either. Moreover special attention will need to be given to making sure that samples from depth are not contaminated by Martian surface materials and dust as they are transferred to the return canister.

At the Earth end of the transfer, there will be a need for quarantine not only to deal with potential human pathogens, but also to prevent contamination of the terrestrial environment by pathogens that have a potential of being inimical to other terrestrial life forms including plants. The analytical laboratory facility for science should, indeed, be in the same location as the quarantine facility; all under one protective roof, as it were.

While the possibility of contaminating the terrestrial environment by organisms from Mars is scientifically very remote, considerable effort will have to be made to educate the public, especially some of the most vociferous elements, on this fact. In view of the recent (1989) unfounded and uncalled for commotion concerning the use of radioisotope thermoelectric generators in *Galileo,* there will undoubtedly be a great need for educational programs through the media to provide basic information to the general public before samples can be returned from Mars.

Scientists are actively engaged in defining rules for planetary protection through the medium of intensive workshop sessions. One was held in Washington, D.C. in June 1988 and another proceeded further at a workshop held in Palo Alto, California in March 1990. These workshops benefited from the knowledge and experience of specialists in the various disciplines involved; science, law, engineering, public relations and communications. We can rest assured that by the time the U.S. is ready with the funding for a Mars sample return, the procedures for guarding against contamination of Mars and Earth will be firmly established as a result of these workshops.

Looking further ahead some planners have resurrected the older missions suggested for *Pioneer*-class spacecraft which were discussed earlier in this chapter. A Mars aeronomy explorer mission has been suggested for the 1990s to expand our knowledge about the upper atmosphere and ionosphere of the planet. This is similar to the proposal made much earlier for this type of mission to be accomplished by a *Pioneer*-type spacecraft. The spacecraft would be placed in an elliptical orbit and equipped with propulsion so that the periapsis could dip into the atmosphere and make in-situ measurements, with the propulsion system compensating for atmospheric drag from time to time as needed.

Such a spacecraft should ideally be kept in orbit around the planet for at least one Martian year. It would be used to determine how the structure of the upper atmosphere and the ionosphere changes on a seasonal and daily basis. It would investigate the interaction of the solar wind with the Martian ionosphere and what manner of protection might be given by the weak magnetic field of Mars. It would also measure the rate at which atmospheric gases are escaping from the atmosphere into space and would thereby provide information from which could be derived a better history of the atmosphere and how it was changed since the planet formed.

Much of this information is needed for extended missions to Mars in which aerobraking will be necessary to conserve propellants. This is especially so for the Mars Rover and Sample Return mission and subsequent manned missions.

Another proposed mission is a Mars Network Observer, again using a larger and more complex spacecraft than the proposed *Pioneer*-type mission described earlier in this chapter. This mission would have the objectives of placing on the Martian surface a network of seismic and meteorological instruments that could operate unattended for many years. The instruments would be carried to the surface by penetrators generally and at a few specific sites by soft landers. These stations would reveal a great deal about the current volcanic or tectonic activity on Mars and allow a modeling of the interior of the planet. The meteorological instruments would, of course, provide a planetwide record of weather conditions.

Neither of these missions had been authorized as part of the U.S. space program at the time of writing, despite lengthy discussions about them over many years.

There are two important scenarios for the further exploration of Mars which are now becoming quite apparent. In one the United States takes advantage of the Soviet invitation to collaborate in the Russian exploration of Mars with perhaps some competition in certain parts of the exploration. In the other the United States adapts human exploration of Mars as a long-term space goal of which a Mars sample return mission is an essential precursor to demonstrate a number of technologies including aerocapture, orbital rendezvous at Mars, and landing, and to obtain the best possible information on the Martian environment. It is no longer a question of whether a sample can be returned from Mars, but when and who will bring it back.

The United States has an opportunity to offer world leadership in space, by offering to manage an international endeavor to develop permanent space settlements and encouraging worldwide participation to stimulate global economy. It is disgusting that 13 million children died of starvation in 1988 while nations of the world spent enormous sums ostensibly defending themselves against themselves. If these nations had spent similar amounts on an international long-range program to expand human activities into the Solar System, the economic multipliers could have been enormous and much poverty and its resultant hunger might have been avoided. We need to encourage other nations to participate in developing the industrial base to participate in producing components for advanced spacecraft. We need to provide production loans and know-how for peaceful technology rather than loans to buy our weapon systems. We need to encourage other countries to become involved in the design and participate in the construction of extraterrestrial and international bases, ultimately to undertake major extraterrestrial activities such as mining the asteroids. The *Apollo* program of an earlier time demonstrated the economic multiplier when it boosted the U.S. economy. It led to the building of new cities, produced new advanced technologies and many technology spin-offs, encouraged young minds to tackle very difficult problems in achieving the "impossible," founded new industries with new manufacturing techniques, and kept people happy in creative productivity working toward a clearly defined goal.

This human productivity could be stimulated on a global basis by good national and international leadership. But it needs people with unusual vision, people with a belief in an optimistic future for humankind freed from self-imposed limitations, and freed from invented enemies. Perhaps we shall see emerge from among the children of the space age, that group which represents half the people on Earth today born since the first men set foot on the Moon, new leaders with such supraglobal vision. The *Apollo* program, which was the first advance into space, came from a wave of creative immigrants from Europe who wanted to apply their technology into a new dimension and found ready partners in captured German rocket engineers and the native Americans who also had developed creative technology to meet the demands of World War II. Perhaps the next impetus to move into the interplanetary age will come from the new wave of technical specialists, even perhaps from the influx of Asiatics, who are already proving to be innovators and exemplary students in the natural sciences and in advanced engineering and computer sciences. Perhaps a new group of children of the space age will create a plan for a stimulated global society able to lift humankind from the restrictive confines of Earth and at the same time lift many nations from the impossible burden of international debt.

In the next chapter we will look at the beginning of the new thrust into space, the establishment of a permanent human presence on Mars, and some of its long-term implications for the future development of our species.

On July 20, 1989, on the occasion of the twentieth anniversary of Apollo 11, President George Bush announced a national goal of establishing the United States as the preeminent spacefaring nation with a commitment of a sustained program of manned exploration of the Solar System and the permanent settlement of space. First objective for the 1990s is space station *Freedom,* then early in the next century, back to the Moon, and this time to stay. "And then—a journey into tomorrow—a journey to another planet—a manned mission to Mars." On November 2, 1989, the President approved a national space policy to update and reaffirm U.S. goals and activities in space. NASA responded with a 90-day study of the main elements of an Exploration Initiative. This was published on November 20, 1989.

The study consisted of analysis of technical scenarios, science opportunities, required technologies, international considerations, institutional strengths and needs, and estimates of resources required. NASA started to work closely with the National Space Council staff to structure a nationwide program seeking technical innovations and new ideas to fulfill this commitment to space activity. Critical technologies, new types of space vehicles, and different approaches to implementing the space exploration initiative are being investigated.

On February 21, 1990 a memo from the White House confirmed the Moon and Mars initiative as a technology program for several years with no immediate decision on the mission architecture but with at least two alternatives to be defined. If the nation and Congress support this program, America indeed will be awake once more to the enormous opportunities offered by the Solar System to unite humanity and to develop the human species to a new level of broadened consciousness.

6

MARS OUTPOSTS, BASES, AND SETTLEMENTS

In the early 1960s NASA contracted for more than sixty studies of manned flights to Mars. The technology of *Apollo* was adequate to have sent manned expeditions to Mars using rotation to simulate gravity and avoid zero-gravity biomedical problems, assembly of the spaceships in Earth orbit, and relying on the tremendous lifting capacity of the mighty Saturn booster. Many of these advanced launch vehicles developed by Wernher Von Braun's team were ultimately left around to deteriorate without any chance of being used in space, a direct consequence of the shortsighted decision to have only one means of sending payloads into space—the Space Shuttle transportation system.

The administration of that era in the late 1960s suggested that the United States should continue its space exploration success in reaching the Moon with a manned landing on Mars by the end of the century. This would have been quite feasible using Saturn V boosters. It would have relied upon sending two command module-type spaceships with a total crew of twelve astronauts on a 600-day mission to the Red Planet. The command ships would have been assembled in Earth orbit from modules transported there by the Saturn Vs. A Mars excursion module developed from the Lunar Excursion Module technology would have landed on Mars, while the return interplanetary spaceship was left in orbit for a later rendezvous, as was done with the *Apollo* Command Module in lunar orbit. At that time von Braun, who for over two decades had been pushing for flight to Mars as outlined in the book *The Exploration of Mars* which he co-authored with Willy Ley in 1956, argued that the mission would be no greater challenge technically than committing to land on the Moon was in 1961.

Even earlier, in 1952, the author published a concept for a nuclear-propelled Mars mission (figure 6.1) using assembly in low Earth orbit by means of winged shuttle-type ascent/descent vehicles. The interplanetary vehicle, assembled in modules, was intended to carry a descent and ascent stage for transportation between Mars satellite orbit and the Martian surface. Liquid hydrogen was to be the main interplanetary propellant, heated by a nuclear power plant of the type later being developed in the United States by the Nerva program of the early 1960s.

When the author visited Jackass Flats, Nevada, in 1969, the NERVA program was about to be canceled. NERVA was an acronym for nuclear engine for rocket vehicle application. A joint program of the Atomic Energy Commission and NASA had completed a long series of tests during which the engines had performed safely. Some 12 consecutive reactors and engines had accumulated over 14 hours of power operations, four of which had been at high power. L. J. Carton, manager of advanced projects at Aerojet, builders of the engine, said in an interview at that time that the NERVA engine was intended to be used in missions to Mars. Test engines had been operated at 55,000 pounds thrust and were capable of being throttled and restarted. Approximately 10 feet in diameter and 34 feet in length, the planned NERVA flight engine would have weighed 20,000 pounds including 3000 pounds of radiation shielding, and would have developed a thrust of 200,000 pounds and a specific impulse of 825 seconds, far ahead of any chemical rocket. The aim was to have a flight engine tested and operational by 1977 for the United States to explore Mars before the end of the century. But despite the successful tests the program was canceled.

In the 1960s there were many proposals for manned missions to fly by Mars and Venus and to land on Mars. Early in the 1960s, for example, Robert L. Sohn of TRW Systems presented a paper on future manned planetary missions. In it he pointed out that there was a growing concern and sense of urgency about developing post-*Apollo* plans. The massive buildup in equipment, facilities, industrial base, and manpower to complete the *Apollo* mission had created a fantastic mission capability that would be lost unless new manned space goals were established quickly. How right he proved to be. The *Apollo* capabilities were allowed to disintegrate for lack of national leadership in space planning.

Sohn proposed several manned missions based on *Apollo* technology. In one of these the mission would have occupied 435 days with 20 days in orbit around Mars during which the four-man crew would have sent an unmanned lander down to the planet's surface. The lander would have gathered a sample and returned it to the orbiting manned spacecraft. The mission would have used three Saturn V launch vehicles to assemble the interplanetary spacecraft in low Earth orbit. That spacecraft would have been a modified Saturn upper stage enclosing a modified *Apollo* command module. The crew would be able to move along a central cylinder to a command center and "storm cellar" where they would be protected from solar storms which might be encountered during the long voyage. Aerodynamic braking was proposed for use at Mars and on the return to Earth.

In 1967 NASA headquarters had an enthusiastic team studying the manned Mars mission. Edward Z. Gray, Director of the Advanced Manned Missions Program, presented details of some of the missions studied when he gave a paper to the American Astronautical Society at the Fifth Goddard Memorial Symposium in March 1967. It was shown that based on *Apollo* technology, manned missions could be sent to Mars in 1975, 1977, and 1979. The general arrangement of the manned flyby spacecraft was to

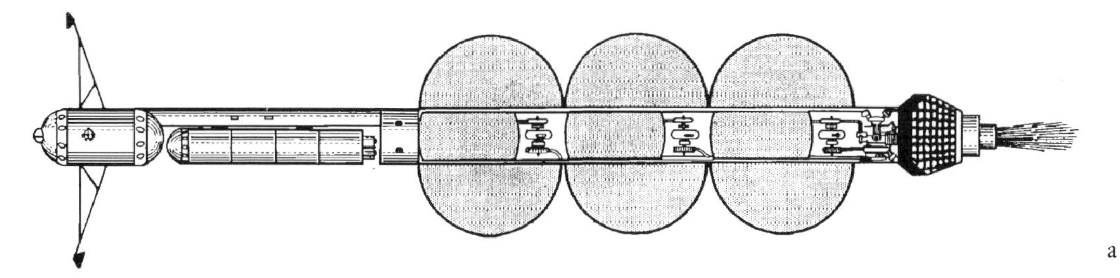

a

b

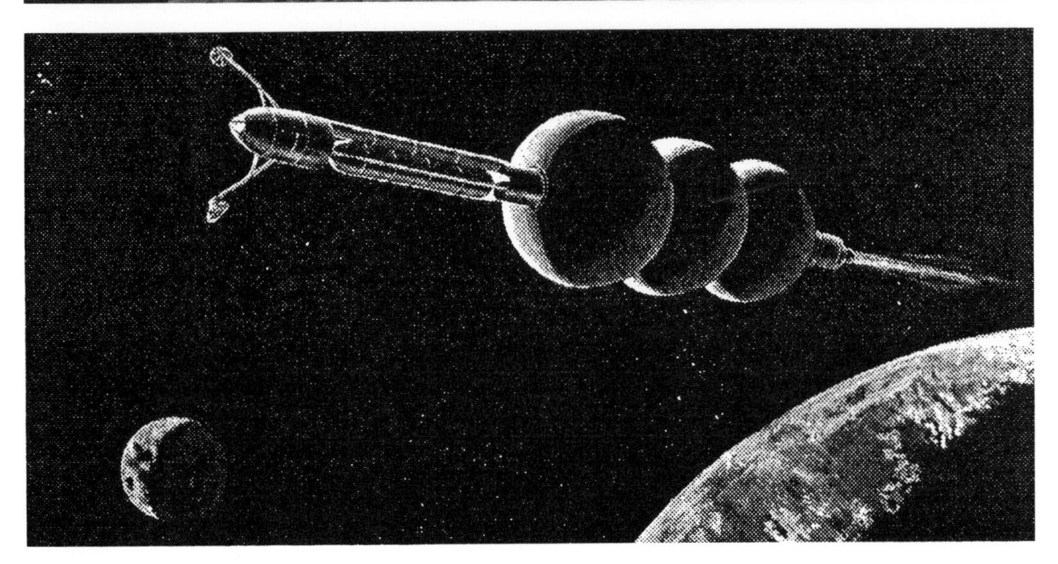

c

FIGURE 6.1: a) This schematic diagram of a nuclear-powered spacecraft for a manned mission to Mars first appeared in the author's book *Rocket Propulsion* published by Chapman and Hall in 1952. A descent/ascent vehicle was carried on a long boom separating the crew compartment from the propellant tanks and the nuclear/hydrogen rocket engine. b) The Mars spacecraft was intended to be assembled in orbit using winged shuttles to ferry materials and crew from Earth's surface to low Earth orbit, as shown in this illustration from the author's *Rockets and Spaceflight* published by Hodder and Stoughton in 1956. c) In this painting by the author, the Mars spacecraft is shown leaving Earth orbit with Earth and Moon in the background. This illustration was also first published in *Rockets and Spaceflight* in 1956.

use a modified Saturn upper stage. The Mars manned spacecraft would carry a biological laboratory for the crew to analyze samples brought up from the surface by unmanned vehicles. The same paper had contributions from a co-author, Franklin P. Dixon, Director of NASA's Planetary Mission Studies, proposing unmanned camera equipped orbiters, geophysical landers, and surface sample return spacecraft. The paper also included plans for a manned landing on Mars based on an uprated Saturn V launch vehicle with strapped-on boosters to carry modules for assembly in low Earth orbit, a Mars excursion module to land a crew on the Martian surface and return it later to orbit, and an Earth return module. The interplanetary spacecraft was configured somewhat similarly to the author's 1952 study. Essentially the crew-carrying part was at the front followed by propellant tanks and propulsive engines for leaving Mars, arriving at Mars, and leaving Earth. The leaving Earth stage was the largest stage and it was located at the tail end of the interplanetary assemblage.

In January 1967, the Space and Information Systems Division of North American Aviation, prepared a detailed briefing based on studies over a period of four years on manned missions following *Apollo,* including a mission to Mars. Once again it was shown that several launches of the Saturn V big booster would suffice to assemble the interplanetary mission in low Earth orbit. The SISD study gave details of various missions using fly bys and aerobraking, and diagrammed a spacecraft design which would use nuclear engines to leave Earth orbit and inject into the trajectory to Mars. The Earth entry module was a modified *Apollo* command module; the Mars mission module which would orbit Mars was a much larger new unit. Also, the spacecraft carried a Mars excursion module to land the crew on Mars and later return them to the orbiting spacecraft. The overall configuration was much the same as that of the earlier proposals. The design included a capability of separating the interplanetary spacecraft after leaving Earth orbit. The two sections would be connected by cables so that gravity could be simulated in the manned modules by spinning the assemblage at about four revolutions per minute. The author interviewed a number of the engineers working on this program at NASA headquarters and at aerospace companies and was able to review the detailed reports and engineering plans. He was impressed with the tremendous amount of detail that had been amassed to demonstrate that a manned mission could be sent to the Red Planet in the decade following the *Apollo* landing on the Moon. The technology was available; national leadership was not.

In the late 1960s Congress and the press were negative about manned missions to Mars. Basically the unpopular Vietnam war of advanced military technology waged against an idealogical enemy on the far side of this world was antagonizing the public against science and technology in general and was squandering tax dollars at such an unprecedented rate to make major nonmilitary projects economically unfeasible.

It was distressing to visit the aerospace companies that had planned the manned missions to Mars and a wide program of unmanned exploration of the Solar System and find that the engineers had been laid off, the great parking lots were empty, the momentum of America's move into space had been lost. Great Saturn boosters created by Von Braun's Huntsville team were deteriorating in storage, rocket test stands were rusting at propulsion test sites, the nuclear rocket program had been canceled. Man on Mars became relegated to something to be achieved by our descendants in the next century. Or so it seemed for a decade or more. Then it became increasingly evident that the Russians were not accepting this limiting viewpoint regarding manned exploration of the Solar System. They continued to develop bigger launch capabilities into a whole fleet of launch vehicles. A Soviet continued human presence in space, first in Earth orbit and later on other worlds, become patently obvious to any knowledgeable observer of the international space community.

The result was that a call for a return to Mars with a manned mission to the planet as a goal began to be heard in the 1980s at many meetings in the United States. This manned mission to Mars is today visualized in many ways. Current thought assumes assembly in orbit using the Space Station, *Freedom,* as a base for construction of the interplanetary spaceship in Earth Orbit. The Mars cruise spacecraft would then be accelerated by an interplanetary transfer vehicle, also assembled in orbit, from low Earth orbit onto its interplanetary trajectory. The cruise spacecraft carrying vehicles for a Mars landing would use aerobraking at Mars to enter an elliptical orbit around the Red Planet. The landers would be detached and descend to the surface using rockets and aerodynamic braking. After a period of surface exploration, a spacecraft would return to orbit and rendezvous with the interplanetary vehicle. After transfer of the crew, the interplanetary cruise vehicle would return to Earth and use aerocapture there to enter a low Earth orbit.

There are three stages in the manned exploration of Mars. The first is one of exploring individual sites to find the best locations for the next stage of exploration; the setting up of outposts on Mars. When experience has been gained with the outposts, the final stage can be entered: establishing a permanent base or system of bases on Mars.

The general objectives of manned missions to the Red Planet are to explore the surface of Mars at many sites to develop planetary science in greater depth. The crew that lands on Mars must be able to live off the land as did the early explorers of Earth's continents. Manned missions to Mars will be a major human step into the universe. They will demonstrate the feasibility of our species developing into interplanetary man and of the survival of humans in deep space. Intellectually they will provide an entirely new viewpoint of our planet and its life as the Mars colonists look back on their own world and see it as merely one of many bright stars in skies of the Red Planet. If the colonists were to develop a mythology they would be more justified in naming Terra the god of war than we war-plagued terrestrials were in giving Mars that distinction!

Science objectives for the manned mission include a deeper understanding of human physiology and of human psychology and sociology in isolation such as never before encountered. Also we shall gain understanding of human reactions in a confined environment for unprecedented periods, especially how humans can deal with the stresses, mental and physical, caused by that environment. There will be great benefits in the physical sciences too, in astronomy, geology, meteorology, biology, and in applications technology of using the space environment in many ways, including manufacturing. Geology on Mars is much richer than on Earth's Moon. Astronauts in orbit can drive unmanned rovers to distant parts of Mars. These rovers will allow us to learn much more about Martian geology, about life on Mars, and the planet's atmosphere. From orbit, too, the diminutive satellites of Mars can be explored.

The configuration of the first manned spacecraft that lands on Mars must include a Mars ascent vehicle, ascent propellant, probably two rover vehicles, a power system consisting of solar cells or a small nuclear-powered generator, a ramp to gain access to the surface by the rovers, and a human habitat. Figure 6.2 shows an artist's concept of a manned spacecraft leaving Mars after the crew have made an initial exploration of the planet. Note the ramp for the roving vehicle they used in their exploration.

FIGURE 6.2: Artist's concept of a manned spacecraft leaving Mars after the crew have completed an initial exploration of the surface. A ramp extends on the right from the part of the vehicle which remains on Mars. This was the ramp along which the rover vehicle gained access to the surface. (Martin Marietta)

An idea of the magnitude of the task of sending a manned mission to Mars is gained by noting that for every pound of payload to orbit around Mars there will have to be several shuttle flights to Earth orbit. The existing Space Shuttle is inadequate. A heavy lift launch vehicle must be developed, but as yet there are no nationally approved programs to do this.

Extravehicular activity on Mars will require that astronauts have lightweight pressure suits (figure 6.3) suitable for Mars' three-eighths of Earth's surface gravity. A hard suit will weigh about 175 lb (80kg), and provide for use up to eight hours before it needs to be refurbished. Its characteristics will be to have wrist and neck seals, good thermal balance in the extreme cold Martian climate, easy recharge of its life support system, and to be easy to reuse; that is, it will require only minimum refurbishing after sorties on the Martian surface. A suitport concept, similar to that proposed for use with the U.S. Space Station, *Freedom,* appears to be mandatory to avoid bringing the suit and Martian dust into the habitat or into a roving vehicle and to prevent environmental contamination of the habitat or vehicles.

The Moon rover used by *Apollo* astronauts (figure 6.4) was a "bare bones" vehicle which was open to the lunar environment and required that the astronauts wear spacesuits. The lunar rover could only go safely a distance equal to that over which the astronauts could walk back to the Lunar Excursion Module if the rover propulsion should have failed. This distance was a few kilometers only. On Mars there is a need to go much further in the exploration, either by using a more advanced roving vehicle or by relying on remotely controlled unmanned vehicles instead of manned rovers.

A meaningful exploration of Mars requires that the explorers should be able to travel 60 miles or more from their base habitat. This, in turn, requires use of a more complex rover with adequate protection for the crew of two people and a total emergency transportation capacity for four people. There should be at least as much room in the rover as there was in the capsule of the *Gemini* spacecraft, and the rover's environment should permit the crew to operate within the vehicle without spacesuits and should provide for them to don spacesuits for activities outside. Also, there should be two rovers to provide a rescue capability if one should fail.

FIGURE 6.3: The atmosphere of Mars is not sufficient for humans to operate on the planet's surface without protective suits. In this artist's concept two crew members are looking across the surface toward their rover which is partially obscured by a dust storm. (NASA-Johnson)

FIGURE 6.4: Invaluable experience in operating vehicles on another world was gained with the Moon rovers used in the *Apollo* program. Mars rovers will need to be much more sophisticated and able to traverse much greater distances, preferably allowing the crew to operate them without having to wear restrictive spacesuits, and possessing rescue capabilities in case of breakdown of a vehicle. (NASA-Johnson)

A great benefit from humans on Mars will be derived when drilling for core samples as *Apollo* astronauts did on the Moon (figure 6.5). If we are restricted to robotics and unmanned rovers, core samples might only be possible to a depth of a few meters, which is not sufficient to gain a good understanding of the layering and composition of the Martian regolith. By contrast, humans using portable equipment can drill to many meters if necessary. This is important in the search for water beneath the surface. Once a water source is found on Mars, the way is opened for a permanent base to be established at that site. With plentiful water, plants could be grown in the habitat to support the base indefinitely.

Mars has many resources available for use by inhabitants of a Martian base; for example, water, oxygen, nitrogen, and carbon dioxide. Some of these resources are useful to provide life support to the crew and to make plant food. Others can be used

FIGURE 6.5: Humans are most important when making experiments on other worlds. This was proved in the *Apollo* missions to the Moon during which astronauts were able to collect interesting and important samples that could have been missed by automated or semi-automated machines. Astronauts were also able to take deep core samples more easily than could any machine. A human presence on Mars is essential for a meaningful search for extinct or extant life upon the planet. Moreover, the operations on Mars must be spread over much longer time than was possible on the Moon with the individual *Apollo* missions. (NASA-Johnson)

to make rover, or rocket, or aircraft propellants. For example carbon monoxide and oxygen propellants can be made from carbon dioxide, and hydrogen peroxide propellant can be made from water. Also there are metals available on Mars. The Martian soil can be used to blanket habitats and protect the crew from solar protons and high energy cosmic radiation.

In chapter 5 it was shown that precursor missions to the manned mission are the Mars orbiter, an atmosphere in-situ explorer such as *Pioneer* Venus orbiter or the Earth *Atmosphere Explorer* satellites to gather data on the Martian atmosphere regarding aerocapture techniques, the Mars Rover and Sample Return mission, Shuttle experiments, and manned experience and space building experience gained in the *Freedom* Space Station. For example, combinations of Martian dust with moisture may lead to a corrosive situation on suits and rovers. This must be resolved in advance by a sample return mission. Also, living in the closed ecosystem of a space habitat can be experienced in the Space Station *Freedom* (figure 6.6) in preparation for the long interplanetary voyage and the period that must be endured isolated on Mars.

To send a manned mission to Mars requires many supporting tasks, much like the *Apollo* missions to the Moon but on a much grander and more far-reaching scale. To begin with, the project will need new mainline launch vehicles including a high lift vehicle to carry substantial payloads into low Earth orbit. A new propulsive system to move the Mars craft from Earth orbit into the interplanetary transfer orbit is needed. Considerable work needs to be done on space medicine to ensure that the astronauts remain healthy during their long absence in space and on Mars. This applies both to their physical and mental well-being.

Mars mission planners have to study the social and psychological factors relevant to the design of the interplanetary vehicle and the Martian base. For example, there is a need to investigate how public areas should be arranged, how color schemes should be selected for crew quarters, how to ensure privacy in the restricted space available, and how open passageways and interior design arts can be used to create illusions of

FIGURE 6.6: Living in the closed ecosystem of a space habitat can be experienced in the modules of a space station. Located from the top right of the grid structure is a logistics resupply module containing perishables, a habitat module or living module positioned vertically, a life sciences laboratory module shown horizontally, a materials processing manufacturing and technology module shown vertically at left, another habitat or living module shown horizontally across the horizon, a small potential pharmaceutical equipment package, and the rear of a hanger for fuel supply tanks. (Boeing Aerospace Company)

spaciousness. Crew members on a Mars expedition will be asked to perform risky and demanding jobs while living in confined and isolated environments away from Earth. While the United States has virtually no practical experience in space on such matters, analogs are available. Psychologists have been studying crews of nuclear submarines, personnel at Antarctic outposts, and North Sea diving teams who work for several weeks in a pressurized environment 600 feet (200 m) underwater.

There may also be hazards that cause injuries to crews in space and on Mars. Since return to medical facilities on Earth will be impossible, inflight capabilities must be designed. NASA has already approached this problem in connection with the *Freedom* Space Station and has designed a health maintenance facility for use in that station. This type of facility might also be taken to Mars. It is designed to permit preventive, diagnostic and therapeutic care, applying state-of-the-art medical capabilities and hardware to the specialized and self-contained spacecraft environment. It includes technology for cardiopulmonary resuscitation (CPR), a portable minor surgical module with sterilized surgical supplies, anesthetics, restraint systems, electrocautery, and specialized equipment to work under the conditions of interplanetary flight.

Research needs to be done on the optimum size of crews for the mission; one suggestion has been for a crew of seven. Other research is needed on how propellants for the return to Earth can be stored in orbit around Mars during the exploration of the Martian surface which may last for several months. Considerable effort is needed to design the rover vehicles, and systems for aerobraking and for descent to the surface of Mars.

There are several viable strategies for planning and implementing the manned exploration of the Red Planet. Of two considered here, the first is most undesirable but most likely under the current system of funding and planning space activities. It is one of sporadic, semi-random evolution, a continuance of the present muddling U.S. process without top political leadership or any clearly defined long-range and continuously funded plan, and without any identified milestones by which progress toward a preset goal can be measured.

A second strategy is a more desirable one of directed evolution. This would be based on a well thought-out and implemented long-range plan aimed at a clearly defined goal; a focused program like the *Apollo* program to reach the Moon. Such a plan has now been developed by NASA and by contractors such as Rockwell International. But the plan needs approval by the public and implementing by a strong national leadership. As contrasted with the *Apollo* program, however, which was defined as one to reach the Moon and return safely to Earth before the decade was out, the Mars program is to reach Mars and commence a permanent presence on and development of that planet as part of human expansion into the Solar System. Merely reaching Mars and returning to Earth is in itself not a desirable goal. A manned mission must lead to further desirable goals if it is to be worthwhile. We must learn from the politics and short-sightedness of the *Apollo* program, which virtually led us nowhere and was comparable to discovering the American continent, taking some samples back home, and deciding to stay home in Europe rather than continuing to develop the new world.

While such a long-range strategic space plan is only in a semi-official exploratory stage in the United States, the Soviets appear to be actively following the desirable strategy. For years they have been building up human experience and technological capabilities with their space stations. As described in an earlier chapter, they started on the precursor mission to Phobos to gather information on that Martian moon which might be a source of propellants. They have a series of missions planned for each launch window in the 1990s which will survey Mars from orbit, land unmanned rovers and other unmanned stations on the Martian surface, and return Martian soil and rock samples to Earth. Then will come the manned missions and establishment of a permanent base on Mars. By the 100th anniversary of the Soviet revolution, the Red planet could have been "The Red Planet." However, the Soviets in 1989 had problems also. The cold war and the expensive Afghanistan debacle led to economic difficulties paralleling those in the United States resulting from Vietnam and a massive defense program. These economic problems will most probably slow the Soviet expansion into space and provide the United States with a new opportunity to lead the world into the Solar System, beginning with permanent bases on the Moon and on Mars.

There are many short-term science advantages of having humans on Mars, and a great amount of grass-roots support has been demonstrated in the United States in the form of the *Case for Mars* conferences that took place in 1981, 1984, and 1987, from which detailed outlines and a mission profile for manned exploration of the Red Planet have been documented in Proceedings of the American Astronautical Society. When a scientific symposium on Mars was held during 1988 in Sunnyvale, California, a brief announcement was made that the public could attend an evening meeting at which some of the scientists would talk about Mars. The crowds that converged on the lecture hall were unbelievable; over a thousand people had to be turned away and there was no parking space free within blocks of the venue. Even some of the scientists were unable to get into the lecture hall.

Although there seems to be no difficulty in obtaining grass-roots support and that of many scientists for a manned mission to Mars, strong presidential leadership is needed to make the mission a reality. Many people are pushing for clearcut objectives in American science; the major objectives of the next few decades include the supercollider for atomic studies, the mapping of the human genome, and the manned expedition to Mars. In many ways these are complementary projects since the supercollider may reveal information about basic physics of matter and energy that can be applied to the

very advanced propulsion systems that are needed for humans to explore the Solar System in detail and go beyond it into interstellar space. Also a knowledge of the human genome may be a requirement if we are to have humans adapt to life on other worlds and to deep space missions such as those to other star systems.

A rationale has been developed for the setting of our priorities in space. This includes the fundamental human craving for national pride and prestige. We see this exemplified in many aspects of our society: to have the best athletes, to have the most successful commercial enterprises, to have the greatest military prowess and weapon systems. Do we also want to lead in the human expansion into the Solar System, or watch other nations go where we originally helped to lead the way to a new dimension for our species?

Spaceflight has many aspects that fit the requirement to stimulate national pride and prestige. It is something that can be accomplished, not a mere dream. It is timely because we have the technology to make it work. It has worth because it will stimulate the economy more than any military arms race. An even greater worth is that it will stimulate many young minds that are now being wasted as they wallow in the doldrums of an uncertain future. A national plan for an expanding space program will raise hopes of an exciting challenge to know more about reality beyond the "bottom line" and the paper-shuffling towers of the cities of today.

Manned planetary exploration of Mars is unique as a human activity. There is an unambiguous goal whose achievement can easily be measured. It is timely because it will transcend the current miasma of terrestrial limitations, it will overcome limits to growth. At the cutting edge of our capabilities at a transition stage of humankind it has been likened to the modern equivalent of the search for the holy grail. It will be a search for the very roots of our life form, not of any culture or race. It can be shown to be extremely worthy, linked to facts of accomplishment as with athletic competitions, rather than to vague idealogies or politics.

We are at a time when the only obvious next step in the physical and mental development of the human species is to take a step off this planet. But, paradoxically, the hiatus from the *Apollo* landing on the Moon to today is longer than the time from Columbus landing in the New World to the first settlers and colonists moving there. Why this delay? The answer is simple. While Columbus and *Apollo* both needed government sponsors for the initial discoveries, subsequent development of the New World could be on a private enterprise basis on Earth. By contrast both the initial steps and subsequent major developments in space are restricted to governments. Today, humankind suffers from being overgoverned by people with myopic vision of the future, by people who concentrate on fixing effects rather than dealing with causes. In the United States, for example, one rarely hears at a local, state, or national level any clearly defined plan for a future that will stimulate the people. The United States suffers from a "written constitution syndrome"; once a rule has been written down it cannot easily be changed or should not be changed! Fortunately all human rules and regulations and plans can be changed if we wish to do so. They probably should nearly always be changed because we have proved to be very inept at predicting the future. Nevertheless, a plan for the future that most likely will need to be changed would be much better than no long-range plan at all. Today we have to suffer a pitiful short-term "wish list" geared toward ensuring success in the next election.

There is also, as mentioned in the preface, a serious education problem in the United States, a very different situation from the heyday of the space age. In the 1950s and early 1960s there was an excitement to learn as we thrust into space. Young and old wanted to know more about the planets, more about basic sciences and technologies, more about how to manage large projects to get things done on time and within cost and to make things work reliably and efficiently. Today there is apathy and a general retreat from reality into a fantasy world created by involuntary "brainwashing" of shallow television programming and voluntary chemical pollution of the human biological and neurological systems. We need a vigorous active space program to revive an enthusiasm for education and stimulate a return to reality. Talk to most schools about space exploration and ask the youngsters who among them wants to be an astronaut. You receive an overwhelming show of hands and feel a deep sense of sorrow that this youthful enthusiasm of the children of the space age must be stiffled in disappointment. There are no plans whereby they might be able to journey through space to other worlds. Yet it would seem to be logical that people of all ages who can be shown a challenging and creative future will again become enthusiastic about going to school to learn. Nowadays a large proportion of the population is not motivated. Many people can look forward only to paper shuffling or service jobs, so they are indifferent. There is no challenge and no incentive to work hard when learning to serve hamburgers or when processing interminable paperwork in government or commercial bureaucracies.

Unfortunately there appears to be no longer national pride in the United States. Yet overseas cosmonauts from many countries are all heroes, and the Soviets, despite many economic problems, have encouraged many foreign cosmonauts to experience the excitement and new viewpoint of living and working in space. A human element in spaceflight makes a big difference to the enthusiasm of people for space exploration. As mentioned earlier in this book, in third world countries everyone seems proudly to display a picture of their cosmonaut alongside other national idols.

The country that initiates cooperation and brings partners into its space program is rewarded with the international prestige and acknowledgment as a leader. Today is a very different ball game from the era of the space race. Now the laurels of history and the accolade of the present will go to the nation which leads all mankind into the new era of peaceful space travel.

It is important to understand that there are many economic benefits to be gained from space exploration. Spaceflight is an economic multiplier. This was proved by a number of studies published in the 1960s, before the Vietnam debacle clouded the issue. The economic multiplier of space exploration is a good current reason for undertaking the manned mission to Mars because the U.S. economy needs a good boost if the budget deficit and the astronomical national debt are ever to be corrected.

Making a presentation to the President's Commission on Space in 1985, Carol Stoker of NASA-Ames Research Center said that the next major goal of the U.S. space program should be to establish a permanent base on Mars. Developing a Mars base will drive technology and lead to many innovations. It is achievable within the next 20 to 50 years, she claimed. Groups of interplanetary enthusiasts have long argued for a continued human presence on Mars where people are self-sufficient and rely upon Martian resources for maintaining life support systems (figure 6.7). Equipped with laboratories they would also be engaged in scientific research in the biological, geological, and atmospheric sciences. Mars is an ideal location for a manned base because it is abundantly endowed with all the resources necessary to sustain life. With existing technology the first manned mission to Mars could take place soon after the year 2000

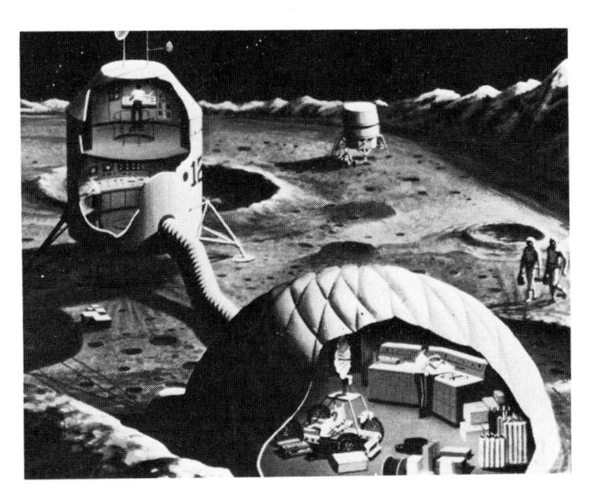

FIGURE 6.7: Many studies and concepts were produced for the establishment of a manned base on the Moon following the *Apollo* program. Unfortunately none materialized. However, much of the work is available for inclusion into the design of a permanent manned base on Mars. This artist's concept shows a typical design for a lunar base that might be adaptable to Mars. (NASA-Johnson)

if implementation of current planning were to start soon. People on Mars are essential to certain tasks such as looking for fossils, and there is a very good chance that Mars will have such fossils from the time when it was a wet and warm planet like the early Earth.

A design for a manned interplanetary vehicle was presented to the Commission by James French, then with the Jet Propulsion Laboratory, during hearings at Stanford University in November 1985. The mission would use three separate vehicles which would be joined together somewhat like a pinwheel so that they could be rotated to simulate gravity during the long interplanetary voyage. Travel time between the two planets could take as much as 30 months overall for a return to Earth. Some Soviet cosmonauts have already spent as long as one year in space under microgravity conditions without serious ill effects. There appear to be no serious reasons why humans should not be able to make long interplanetary journeys, but for convenience sake interplanetary spacecraft can be designed to simulate gravity. This will probably be done so that the crew of a Mars expedition can acclimate to Martian gravity conditions before landing on the Red Planet. Shuttle-type vehicles would be used at either end to transfer people and cargo from the planetary surface to the interplanetary vehicle. This type of Martian expedition has been proposed for nearly 50 years.

Human habitats on Mars would need to be protected with Martian soil to absorb solar protons and high-energy cosmic rays. The atmosphere of 95 percent carbon dioxide can be used to provide oxygen for breathing and for making rocket propellants. Solar energy or nuclear energy could be used as the power source. Ideally a manufacturing plant for propellants should be capable of being deposited on the surface and then function virtually unattended on a continuing basis. Such a unit would extract carbon monoxide and oxygen from the carbon dioxide of the atmosphere, and water from the vapor in the atmosphere. Nitrogen, so essential for plant fertilizer, could also be obtained from the Martian atmosphere.

By using the resources of Mars we can visualize creating a true Martian civilization within 100 to 200 years with people living in protective domes, food growing in Martian greenhouses, and chemical plants producing water and oxygen from the permafrost of the regolith and other necessary chemicals from the Martian soil. However, before we can plan such activities on Mars they must be demonstrated closer to Earth, first in an Earth-orbiting space station and later in manned bases on the Moon. Much research needs to be done on the technology of closed ecological systems, on recycling systems for water, gases, and organic wastes, and on the nutritional and emotional needs for people living under stressful conditions of another world.

How people would spend their leisure time as well as work time would need to be carefully evaluated. Some experience has been gained in submarines, and at bases in Antarctica and the oil "cities" on the North Slope of Alaska (figure 6.8). The latter particularly have demonstrated how self-contained cities can function. They include recreational facilities of many kinds so that people become less aware of the inimical environment outside.

In fact, the building of the oil "cities" on the North Slope is a good analog of establishing a base on another world. The facilities were prefabricated as modules which were barged around Alaska to Prudhoe Bay on the shore of the Arctic Ocean. Each module was a complete section including wiring, ducting, heating, and the like. The Ralph M. Parsons Company developed the concept and was appointed managing contractor for the huge project. It is interesting to note that the modules were mounted above the surface of tundra on piles which were sunk deep into the permafrost in which they were held as firmly as if installed in concrete. This project is an excellent example of the high technology which is now available for prefabricating, transporting, assembling, and bringing into operation completely self-contained human habitats and operating facilities in inimical environments. Such expertise is available for the establishment of bases on other worlds.

Antarctica with its bases is another good analog for Mars. Setting up a planetary simulation facility in a dry valley of Antarctica will be easier and more convenient and a better analogy than setting up a base on the airless Moon or on low-gravity and airless Phobos. An Antarctica base can be used to test equipment to extract water and carbon monoxide and oxygen from the atmosphere, air shell structures and greenhouses, and the self-contained habitats. Most of the equipment required for a Mars base can be tested at an Antarctica simulation base at much less cost than can similar testing be done on other worlds.

The Soviets have already amassed a tremendous amount of experience on living in space. Since the one and only U.S. space station, Skylab, fell from the skies the Soviets have had no less than eight space stations fully operational and manned. There will, indeed, have been a continuing Soviet presence in space for two decades before the United States has another space station if the currently planned schedule for its establishment is not delayed by cost-cutting or further problems with the Space Shuttle. The United States has clearly lost the world leadership in manned space operations that it possessed at the time of *Apollo,* despite its renewed activities with the Space Shuttle transportation system. But the Soviet economic problems may present an opportunity for the United States to regain world leadership.

At present NASA's goals have been stated as including the Space Station, *Freedom,* but there is still lacking a definitive national statement of where that Space Station will lead, or if the goal of "expanding human presence beyond Earth" means establishing a human presence on the Moon and Mars as has been recommended by various study groups, including the President's Commission on Space and the Sally Ride report on Leadership and America's Future in Space. NASA's policy was a little more clearly

a

b

FIGURE 6.8: Experience in establishing bases under hostile environmental conditions has been gained on Earth, particularly in Antarctica and on the North Slope of Alaska. a) This aerial view shows an "oil city" on the North Slope designed and implemented by the Ralph M. Parsons Company. The city was built in modules and transported piecemeal over long distances to the North Slope, very much an analog of what must be done to build and transport to Mars the various modules of a Martian base. b) A team of engineers and managers displays a flag at the North Slope "oil city" during a visit of some journalists to inspect the facility and its operation in isolation from the hostile environment. The author is standing second from the right. (Photos: Ralph M. Parsons Company)

defined in December 1988, when a 51-page report was published entitled *Beyond Earth's Boundaries: Human Exploration in the 21st Century*. The report suggested that the nation should seriously consider four major manned missions: a scientific outpost on the Moon, an experimental station on the Moon to develop technology for humans on Mars, an expedition to the Martian satellite Phobos, and an expedition to the surface of Mars. The report emphasized the need for a firm commitment and early government approval because of the long lead times for such missions.

Assuming the use of technology that is expected to be available early in the twenty-first century, a concept has been developed for a research station on Mars. The concept for the station itself was developed during 1984 as part of the Case for Mars series of conferences and studies. The 1984 workshop was held in Boulder and the concept for the station was detailed in a report prepared by the Jet Propulsion Laboratory and published in 1986. A 1987 workshop, held at the University of Colorado, Boulder, defined the facilities required to assemble an expedition to Mars to sustain a permanent human presence at the research station. There were many sponsors of the Case for Mars activities including the World Space Foundation, the American Astronautical Society, the Jet Propulsion Laboratory, Los Alamos National Laboratory, NASA centers, and the Planetary Society.

The research station on Mars was proposed as a much-needed long-term goal for the space program. A permanent research station was selected rather than a series of separate missions similar to the *Apollo* exploration, which had the Moon as its goal.

A permanent research station offers many advantages that were missed by the *Apollo* program; a greater scientific return for the investment of human effort and economic wealth, greater safety for the astronauts involved, and a potential for eventual expansion into a permanent human settlement on the planet which, in itself, could be an insurance against catastrophic disasters, natural or otherwise, which might end all life here on Earth.

Such a research station (figure 6.9) on the Red Planet would be autonomous and self-sufficient, living off the land as did early settlers from Europe on Earth's newly discovered continents. From our knowledge of Mars it seems feasible to plan for utilizing Martian resources not only to support life but also to provide consumables such as rocket propellants. Carbon dioxide, nitrogen, and water can be obtained from the planet's atmosphere. From carbon dioxide oxygen and carbon monoxide propellants can be extracted, and oxygen for breathable air. Atmospheric nitrogen can be used to produce fertilizers for agriculture.

By organizing overlapping missions to the Red Planet we can maintain a permanent human presence on Mars with all the equipment continually attended. This is similar to the Soviet concept of maintaining its space station for many years with a continuing Soviet presence in low Earth orbit. It contrasts strongly with the American presence on the Moon and the American presence in low Earth orbit which was on a sporadic rather than a sustained basis.

The Case For Mars studies showed that a permanent research station on Mars does not cost more than a series of manned missions to the planet, but has many advantages. If the research station is equipped with rovers, with aircraft, and with propellant manufacturing plants and hydroponic farms, it offers the potential of much wider exploration of Mars and other interplanetary bodies, such as the asteroids.

While a main purpose of a program to explore Mars is science, there are equally, if not more important, economic and humanistic benefits as mentioned earlier.

FIGURE 6.9: Though no concrete plans are underway on a national basis for a manned Mars mission, several studies have been made to investigate such missions that might be made early in the next century following some unmanned missions within the next decade. This artist's concept by Paul Rawlings depicts a permanent base on Mars located in an ancient water eroded canyon near Mars' equator overlooked by the massive shield volcano, Pavonis Mons, in the far background. Various base facilities are illustrated including a rover, a habitation module, power module, greenhouses, the central base, a launch and landing facility, water well pumping station, a maintenance garage, tunneling device, drilling rigs, dish and mast antennas, and a surveillance airplane. (NASA-Johnson)

The scientific mission benefits from the presence of humans because they have unique capabilities that are extremely difficult to automate. Humans can make decisions; they can prepare detailed plans and change these as more information is gathered. They can repair equipment and modify experiments as they progress. They are innovative and able to make improvisations to overcome the unexpected. Humans are also extremely flexible. When coupled with a person's experience, intuition, and intelligence, this flexibility can produce enormous dividends from scientific exploration.

Each time Mars and Earth are positioned favorably for a manned mission, i.e., every two years approximately, spacecraft could fly to Mars with supplies and to rotate crews. To conserve propellant requirements, these missions are envisaged in the Case for Mars studies, as being flybys. The arriving crew of about 15 members and the cargo would be aboard a descent vehicle as the main spacecraft approaches Mars. This shuttle-like winged craft will separate from the interplanetary vehicle and carry the crew and cargo to a landing by aerodynamic braking and rocket power near to the Mars base. The interplanetary vehicle, now unmanned, is programmed to use its rocket engines in a maneuver that will change its trajectory for Earth-return as it flies by Mars. Meanwhile, people from the research base who are returning to Earth (probably less than the 15 new arrivals as the base population is increased) will ascend in a shuttle-type vehicle to rendezvous with the interplanetary vehicle as it is leaving Mars. This shuttle-type vehicle will be carried by the interplanetary vehicle back to Earth orbit for refurbishing. The shuttle on the surface of Mars can be refurbished there for the next Mars-Earth return mission two years later.

On arrival at Earth there are several ways in which the crew can descend to the surface. One is for it to use a small craft to separate from the interplanetary vehicle and rendezvous with a space station. The interplanetary vehicle is aerobraked into low Earth orbit for it to be refurbished for the next mission to Mars. The crew transfers from a space station, such as *Freedom,* to Earth as will other space station crew members by using the space transportation system set up to support that station.

For such a program new spacecraft have to be developed. One of these is the interplanetary vehicle. Another is the Mars shuttle. In addition several support vehicles are required. The Mars shuttle will be of two types: a reusable manned vehicle, and an unmanned cargo ship that only goes one way, i.e., from Earth to Mars and is cannibalized on Mars to make structures at the Martian base.

All the Mars shuttles use aerodynamic braking to slow their approach speed to Mars of between 3.1 and 3.7 miles per second (5 and 6 km/sec) to a speed at which parachutes can be deployed. Each Mars shuttle in this scenario is shaped like a crooked cone with heat-shielded surfaces.

The interplanetary vehicle is envisioned as consisting of three sections that are attached like three spokes of a wheel which is rotated to provide simulated gravity during the long voyage through interplanetary space. While microgravity could be accepted for the *Apollo* missions to the Moon, it is felt that a gravity-like environment will be preferable for the crew's welfare and well-being on the months'-long flight to Mars. The interplanetary vehicle's sections are assembled in orbit close to a space station such as *Freedom.* Each consists of two habitats, based on the space station's habitat modules, attached to a boom and tunnel assembly. This also carries storable propellants, a life-support system, and storage for consumables needed during the interplanetary voyage. These three segments can be connected at one end to form the three-spoke arrangement with the habitats at the ends. Each boom also carries a Mars descent/ascent shuttle. In this scenario each section leaves Earth's orbit separately. One of the supplementary spacecraft is a booster needed to transfer each segment from low Earth orbit to the trans-Mars trajectory. This same booster vehicle is also used to send unmanned cargo shuttles on their way to Mars. The three interplanetary vehicles then rendezvous on the Mars trajectory and when they are docked together they start to rotate. On arrival at Mars the shuttles detach from the interplanetary vehicle and carry astronauts and cargo down to the Martian base.

Critical areas of technology needed for the Mars expeditions have been identified. One of these is aeroassist for which NASA already has a program to test aeroassist fundamentals including the aerothermodynamics of nonequilibrium flow behind the shock front formed ahead of a hypersonic vehicle at high altitudes in Earth's atmosphere. The Aeroassist Flight Experiment mentioned in the previous chapter uses a blunt aeroshell with shuttle-type tiles as a thermal protection system. The test vehicle will be launched from a Space Shuttle in late 1994, accelerated by rocket thrust into the Earth's atmosphere. At an altitude of about 250,000 feet it will be slowed by drag to orbital velocity. Skipping back out of the atmosphere it will be recovered by the Space Shuttle and returned to Earth for evaluation.

This initial experiment is directed toward establishing the basic physical data for the design of a fleet of Aeroassisted Space Transfer Vehicles (ASTVs) to be used when returning payloads from geosynchronous orbit or from the Moon. A major thrust is to validate computer codes for predicting radiative and convective heat transfer during very high speed flight in regions of low atmospheric density. This information cannot be obtained from ground-based tests because only small models can be tested in the ballistic and shock tunnels, and scaling factors are important. Aeroassist is essential for any large-scale development of the Solar System. For example, the amount of material that has to be moved from the Earth's surface to low Earth orbit for sending a manned mission to Mars can be halved if aeroassist is used at Mars and on return to Earth orbit. As was detailed in chapter 5, even unmanned missions to Mars for sample return require aeroassist technology to obtain a reasonable amount of samples.

Another important area of technology is the transfer and storage of cryogenic propellants. It has not yet been demonstrated in space, but is essential for a realistic approach to manned missions to both the Moon and to Mars.

Also, there is a requirement that we should be able to assemble vehicles in space. Some of these vehicles for the manned development of Mars will be quite large, as much as thousands of tons. Assembly might be accomplished by humans, but will undoubtedly rely largely on intelligent robotic systems developed from those now being designed for use with Space Station *Freedom.*

If it is found that humans cannot operate for long periods in space without being exposed to gravity, simulated gravity vehicles will have to be developed. Generally this is assumed to rely upon rotation and centripetal force. The primary engineering challenges are in designing the tethers to hold two parts of the spinning spacecraft, the dynamics and control of flexible structures, and means to despin the assemblage and bring the parts of the spacecraft back together again at Mars and Earth.

For a viable development of Mars a nuclear electric propulsion system appears to be necessary (figure 6.10). This requires a nuclear power source with a specific mass of less than 20 kilograms per kilowatt, an efficient conversion system to change the nuclear energy to electrical energy, efficient thrusters, and means to protect the crew and spacecraft systems from radiation.

Of most importance is the challenge of providing an environment in which crews can work both on interplanetary voyages and on the surfaces of other worlds. This challenge has to be met by designing suits for activities in space and on the surface of Mars and its satellites, life support units, means to monitor radiation and protect from it while crews are in space and on the surfaces of the other worlds, protection from meteorites and from space debris when in Earth orbit, and production of food.

Various scenarios for a manned mission to Mars assume that economy of mass at the beginning of an expedition can be achieved by having the interplanetary vehicles fly by Mars rather than enter an orbit around Mars. The slingshot of Mars' orbital motion and gravity is used with some additional rocket thrust to avoid relying on the more time-consuming Hohmann-type orbits requiring minimum transfer energy. Also, the time crews must remain on Mars before a return journey is shortened considerably. On the negative side, descent to Mars and ascent of the returning crew has to be timed precisely. Other scenarios assume that the interplanetary vehicles will orbit Mars for a period until the planetary conditions are correct for the return journey. This type of

a

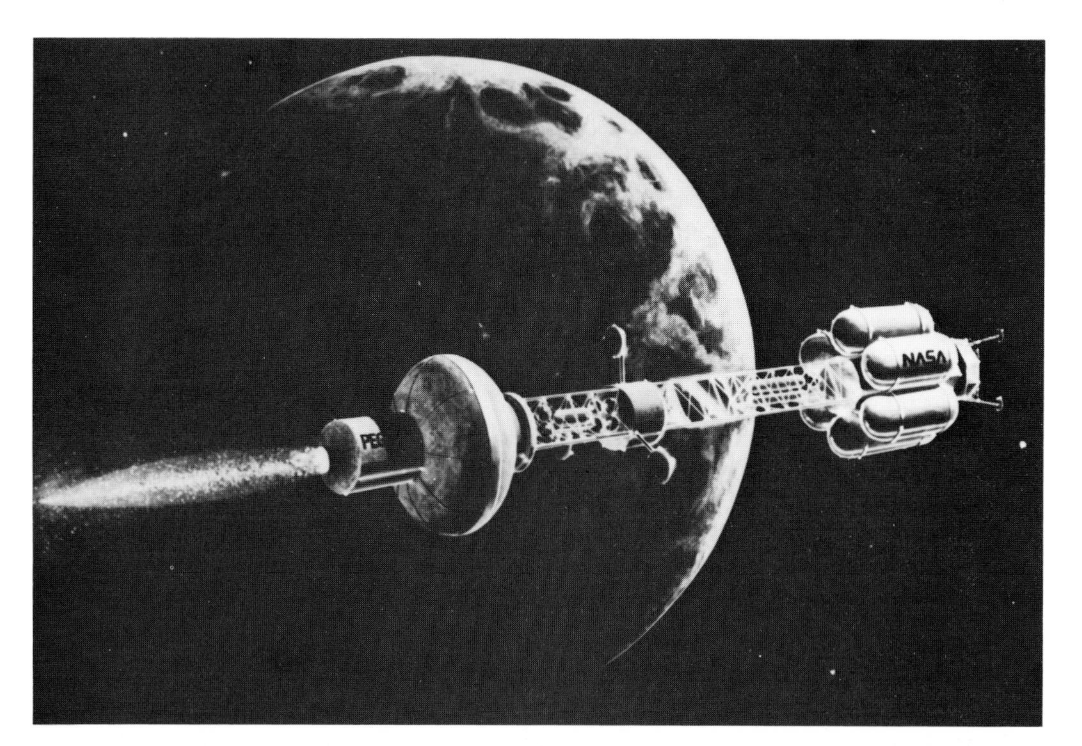

b

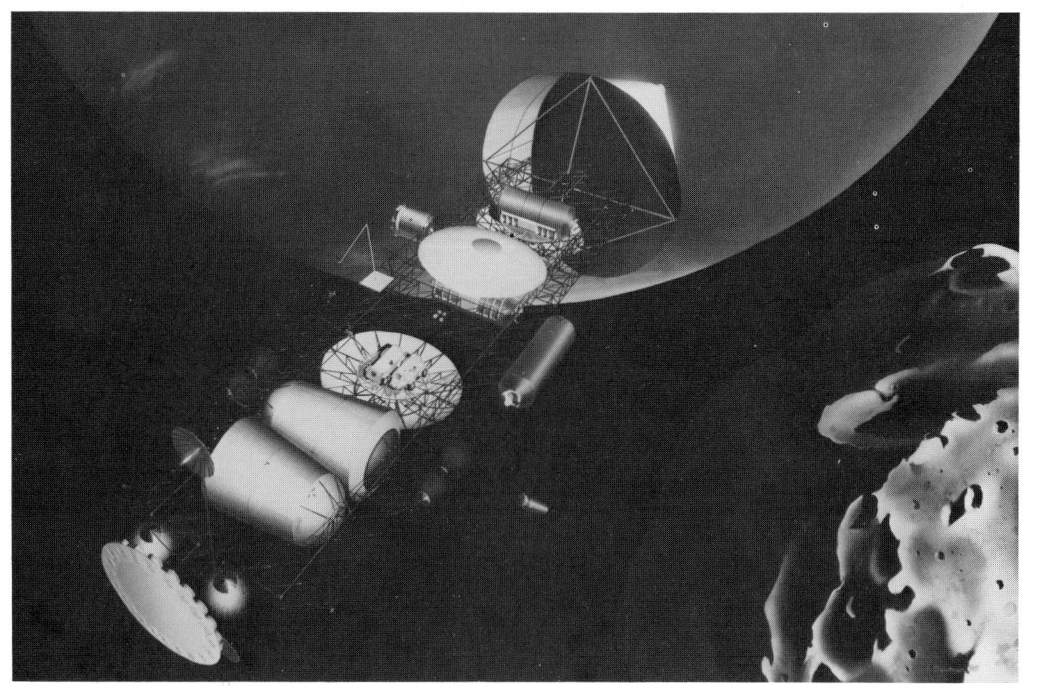

FIGURE 6.10: For economic development of Mars, interplanetary spacecraft need more efficient propulsion systems than can be provided by chemical rockets. a) This 1983 concept of a cargo vehicle relies upon a magnetoplasmadynamic propulsion system using nuclear energy. (NASA-Johnson) b) This 1986 concept shows a nuclear electric propulsion freighter, or space barge, which slowly but steadily carries a massive payload between orbits of Earth and Mars. Launched in advance of a high speed manned flight, the cargo carrier transports most of the equipment and supplies the crew will need on arrival at Mars. This Eagle Engineering concept was referenced by the National Commission on Space. The artwork is by Mark Dowman of Eagle Engineering, Inc., courtesy of NASA (NASA-Johnson)

mission economizes by using low energy transfer orbits, but is expensive in time; it takes much longer for the round trip. Also, it is expensive in propellants needed to enter and leave orbit around Mars. But it does give plenty of time for planning separation of the landing vehicles and for ascent vehicles to rendezvous with the interplanetary vehicle for return to Earth.

For all scenarios the manned expedition must use several clearly defined space vehicles.

A heavy lift launch vehicle is required to assemble materials for the Mars expedition in low Earth orbit. The Soviets have such vehicles, the United States does not. The Space Shuttle is not a suitable vehicle for a manned Mars expedition, it has a low payload for each launch and requires too many flights with excessive launch costs. Also the fleet of Space Shuttles is committed to a tough flight schedule to meet the stated requirements of the military, the Space Station *Freedom,* scientific Earth satellite launches, and the few interplanetary launches now scheduled. It would not be wise to build more of the same limited shuttles which are based on the technology available nearly twenty years ago, since they cannot really do the job either for the Mars project or for establishing bases on the Moon. Major efforts needs to be concentrated on applying the new technologies into the design of advanced and heavy-lift launch vehicles ready for the space requirements of the twenty-first century.

Alternatives include a new super shuttle that is a reusable vehicle, and reusable giant ballistic rockets that can be returned to Earth and reused as are the solid propellant rockets of the current Space Shuttle. Also, the existing Space Shuttle might be modified by the addition of extra solid boosters, the elimination of space allocated to the crew and science teams, and addition of automatic flight control. Such a shuttle might be sent into orbit and returned to Earth automatically, similar to the Soviet shuttle, and would have increased capacity for carrying freight into low Earth orbit. There, materials transported to orbit would be incorporated into the interplanetary and other vehicles required by the mission using a space station as a space habitat for the construction crews (figure 6.11). Heavy-lift launch vehicles have been specified several times in the past, notably in connection with their requirement for establishment of settlements in space at the Langrangian equilibrium points. A typical vehicle would have to be able to lift 150 tons of freight into low Earth orbit compared with the 30 tons payload of the current Space Shuttle. The cost per pound of freight in orbit could be less than half the cost when using the Space Shuttle. Since the heavy-lift shuttle need not be man-rated it can be simplified and its launch costs can thereby be reduced considerably.

Another important space vehicle is that required for transportation between the interplanetary vehicles and the Martian surface. Such a vehicle has been referred to as an aerocapture vehicle since it will rely on atmospheric drag of the Martian atmosphere to slow it sufficiently to be captured by the planet's gravity for a safe entry into that atmosphere using the atmospheric drag again for aerobraking. Moreover, in some scenarios, the aerocapture technique will be used to slow the vehicle from an interplanetary speed to orbital speed around Mars or around Earth.

Two configurations have been proposed for such aerocapture, a ballistic shield (figure 6.12) similar to the *Viking* aeroshell, and an aerodynamic tilted cone similar to the Space Shuttle. The basic requirements are, however, that the vehicle must be able to withstand entry into the atmosphere from a high-speed interplanetary trajectory or from planetary orbit, it should be maneuverable so it can land within a mile or so of a surface base, and it would be advantageous for the landing vehicle to be reusable as an

FIGURE 6.11: The Mars expedition will be assembled in low Earth orbit. This artist's concept by Jack Olsen shows an orbital transfer vehicle being serviced in a hanger of a space station. The hanger serves as a shelter to protect work crews from space debris and provides lighting and easy accessibility for repair and servicing operations. Such facilities will be needed for assembly of the Mars expedition and for servicing and refurbishing interplanetary vehicles and landing/ascent modules on their return to Earth orbit prior to their next mission to Mars. (Boeing Aerospace Company)

FIGURE 6.12: A configuration for a manned Mars lander vehicle is shown in this artist's concept. The ballistic shield extended like a skirt was used to slow the spacecraft on arrival. Another ballistic shield at the top of the ascent vehicle will be used to aerobrake the spacecraft on its return to Earth. This is use of a low lift/drag approach. (NASA-Johnson)

ascent vehicle to carry crew members back to orbit or to rendezvous with an interplanetary flyby vehicle. The aerocapture vehicle should also be of two types: a manned vehicle for transportation of crew members, and an unmanned automatically controlled freight vehicle. While the manned vehicle would need to be refurbished so that it could return crews to space from the Martian surface, the materials of the freight vehicle could be cannibalized to expand the Martian base.

A low drag/high lift aerocapture vehicle suggested by the Case for Mars studies (figure 6.13) would be about 65.5 feet (20 m) in length with a base diameter of about 22 feet (6.8 m). It would weigh about 46 tons when lifted from Earth in a heavy lift vehicle, and its touchdown weight on Mars of 62,000 lb (28,000 kg) would include a payload of about 17,600 lb (8000 kg) carried to Mars. The freight version would land

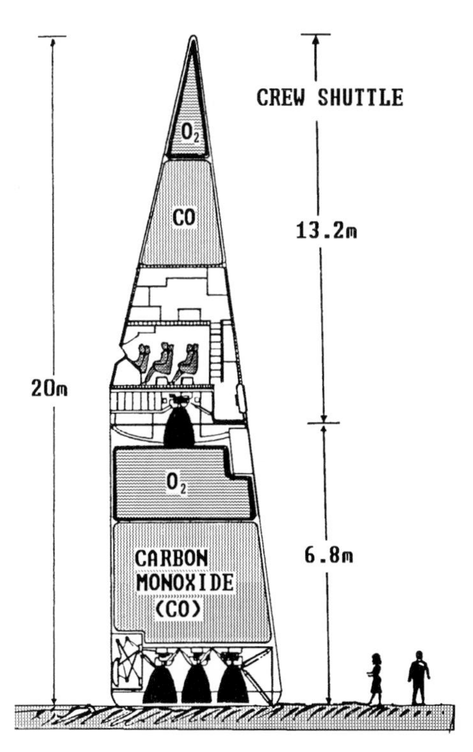

FIGURE 6.13: This drawing shows a type of high lift/drag aerocapture vehicle suggested by The Case for Mars studies. It would weigh about 46 tons and would carry a crew or freight from Mars orbit to the Martian surface and return a crew to interplanetary space for return to Earth. The vehicle would be transported to Mars attached to an interplanetary vehicle.

about 40,000 lb (18,000 kg) on Mars. Since this freight vehicle would not be refurbished it could use a more powerful propellant combination than the carbon dioxide and oxygen combination used for the manned descent/ascent version. Landing of both versions would rely upon aerobraking, a parachute, and retrorockets. The refurbished manned version could carry about 8800 lb (4000 kg) from Mars using a propellant of carbon dioxide and oxygen manufactured at the Mars base from the Martian atmosphere.

An injector spacecraft needs to be designed for transferring Mars-bound vehicles from low Earth orbit into a trajectory that will carry them to Mars. This spacecraft will be assembled and loaded with propellants at a space station. Essentially it is a group of propellant tanks for liquid hydrogen and liquid oxygen and a set of liquid propellant rocket engines. The external tank of the Space Shuttle could be used for this purpose if development costs are to be minimized at the expense of operating costs. These injector spacecraft will be used to boost interplanetary spacecraft on their continuing Martian missions and to boost freight vehicles to Mars ahead of the crews, sending the essential habitats, power generating system, life-support systems, and supplies needed to set up the base initially.

The interplanetary cruise vehicles (figure 6.14) offer many choices. The Case for Mars studies as mentioned earlier have suggested a rotating assembly of three space vehicles connected together at a hub and stretched out like the spokes of a wheel. Rotation of the assemblage during the flight to Mars will provide the crews with a simulated gravity that would make it easier for them to adapt to Martian conditions and

FIGURE 6.14: An interplanetary cruise vehicle is here shown in orbit around Mars close to one of the Martian satellites. (NASA-Johnson)

be agile on arrival at the Red Planet. The first crews especially will need to be very active after landing since they will have to assemble the base from the cargo vehicles, protect it from the radiation environment by covering the habitats with Martian soil, and bring everything into operational status quickly. Each month while in space under microgravity conditions, astronauts and cosmonauts can lose up to 2 percent of their skeletal mass. And as bone mass decreases muscles also shrink, but bacteria appear to thrive. There are reports that after 234 days in space several Soviet cosmonauts were close to losing consciousness on return to Earth's gravity. For seven or more days they were incapable of walking easily. This would be unacceptable for a crew arriving at Mars. It thus seems advisable to simulate gravity during the interplanetary journey.

Each cruise vehicle will consist of a long tubular girdered structure to which are attached propellant tanks, communication antennas, solar power collectors, living quarters, aerobraking shields (needed on return to Earth orbit), and a docking hub section. To this assemblage of approximately 130-feet (40 m) length would be attached a manned aerocapture vehicle. As pointed out earlier, after the assemblages leave low Earth orbit individually and are connected together when on their way to Mars, they are spun by rockets to a rotation rate at which the equivalent of Mars' gravity (approximately 1/3 of Earth's gravity) will be provided in the living quarters. As the interplanetary configuration approaches Mars, it is despun and the crew members transfer to their individual aerocapture vehicles which are detached from the interplanetary assemblage and are aimed into the Martian atmosphere to be captured by aerobraking.

The interplanetary vehicle flies by Mars and returns to Earth where it separates into its three sections that are then aerocaptured into low Earth orbit to be refurbished for the next Mars opportunity. On the second expedition, crew members who have completed their tour of duty on the Red Planet will blast off from Mars in a refurbished aerocapture vehicle and will rendezvous with the fly-by spacecraft and attach their aerocapture vehicles to the arms. They will then rotate the whole assemblage for return to Earth. On approach to Earth the interplanetary vehicle will be despun, the aerocapture vehicles will separate and be individually captured into low Earth orbit. The three sections of the interplanetary vehicle will also separate and be individually aerocaptured into low Earth orbit ready to be refurbished for other crews to be sent to Mars at the next opportunity.

To develop a base on which the future U.S. space program can be planned, NASA's

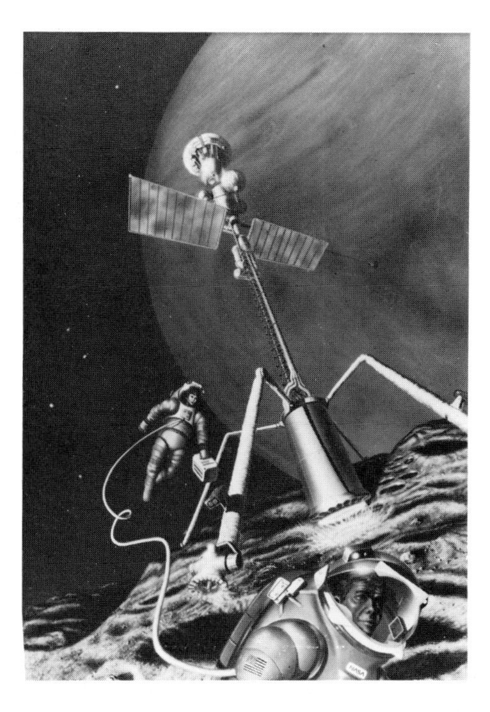

FIGURE 6.15: On Phobos, innermost satellite of Mars and a possible location for extraterrestrial resources, a mobile propellant-production plant is shown lumbering across the irregular surface. Using a nuclear reactor, the large tower melts into the surface. The steam generated is converted into liquid oxygen and liquid hydrogen. This concept was developed by Eagle Engineering, Inc. The artwork is by Patrick Rawlings of Eagle Engineering. (NASA-Johnson)

Office of Exploration selected a number of case studies exploring the technology needed to develop human expansion into the Solar System. Advanced technologies such as aeroassist, propellant production on other worlds, crew habitats, surface power systems, and systems to simulate gravity have been identified and studied.

A preliminary manned mission to Phobos (figure 6.15) has been suggested as a precursor to the Mars landing because it is less complex and would provide the necessary experience of a long spaceflight—many months compared with a few days for the lunar voyages. Also docking with a Martian satellite is fairly easy, and from its surface the crew could direct teleoperated unmanned rovers to explore the Martian surface much more easily than could be done from Earth. Samples of the Martian surface might also be returned to the satellite for analysis there so that an immediate change could be made to the rover's track to gain better samples.

Checking on the satellite itself is also important in trying to determine if these Martian satellites are asteroidal bodies and whether or not they can provide a source of rocket propellants. If they are suitable, they might be used as refuelling stations for the later manned exploration of the planet itself. It is interesting to note that the propellant required to travel from Earth to Phobos and land on that satellite and later return to Earth is less than the propellant required for a landing on Earth's Moon and return to Earth. In fact, providing humans can endure being in space for years, a base on Phobos can be established in much the same way that one could be established on Mars and at considerably less cost. Cargo ship hulls would be landed on Phobos, emptied of their freight and their shells connected by tunnels to form a habitat. Airlocks would give access to the satellite's surface. Then the habitat would be covered with material from the surface to protect the crew from radiation. The result would be a Phobos base of habitats for living and scientific research. This base could be occupied in the future whenever needed, somewhat analogous to the early bases in Antarctica.

NASA's Office of Space Exploration studied a single mission to the Martian satellite Phobos aimed at scaling down the total system needed for an initial human exploration of the Mars system. The objective of the case study was to investigate the practicality of the United States taking leadership by making the first human expedition to Mars, taking advantage of the relative ease of landing on Phobos compared with a landing on Mars, and the possibility that the regolith of the satellite might contain materials needed for continued exploration of the Mars system, including water. A science station would be established on the surface of Phobos from which samples of the satellite's surface could be obtained.

Two types of interplanetary space vehicle would be fabricated on Earth and transferred into Earth orbit by heavy-lift launch vehicles (yet to be developed). A cargo transfer vehicle would be launched in February 2001 ahead of the manned vehicle (figure 6.16). It would carry all the equipment needed to establish the base on Phobos and to land rovers on Mars, and also the propellant needed for the crew to return to Earth orbit. It would arrive at Phobos in October 2002 to await the arrival of the manned vehicle, which would leave Earth in August 2002 taking nine months for the crew of four to arrive at Mars. Two crew members would descend to Phobos and work on the surface. The other two members would remain in the main command vehicle orbiting Mars, from which they would control the robotic exploration of the Martian surface. Propellants for the return journey to Earth, which would take four months, would be transferred from the cargo vehicle to the command vehicle.

In total the crews would spend about one month in the Mars system before returning to Earth, where they would enter Earth's atmosphere for a landing using the *Apollo*-type direct entry in a capsule. With them would be samples of the surface materials of Phobos gathered during the 20-day exploration by two of the crew members, and samples from Mars' surface gathered by the rovers. The crews would have to spend a total of only 14 months in space and on Mars for such a mission. This short time would be achieved by using high-energy transfer orbits made possible because propellants for the return journey would be carried separately in the cargo vehicle.

To produce propellants on Phobos the system should have an output of about 50 tons per year and be fully automated for unattended operation. The facility must be anchored to Phobos automatically when it is landed on the satellite's surface. The gravity of Phobos is too weak for the weight of the facility to be an effective anchor. The system must collect water, believed to be at least 0.1 percent of the material of the regolith and possibly as high as 15 to 20 percent. The water will be driven off from hydrous phases of the regolith material. A thermal reactor with a microgravity fluidized bed will heat the material and drive off the water, which will then be electrolyzed into oxygen and hydrogen. These gases will be liquefied and stored as cryogenic propellants.

Some industrial studies have also been made concerning these technology needs for human expansion into the Solar System. Rockwell International, for example, developed an Integrated Space Plan in 1989 consisting of a long-range systematic projection of evolutionary opportunities for the space program. Its purpose is to assist in the process of synthesizing a space architecture to satisfy space policy objectives. It projects over the next hundred years where we can be in space if the long-range goal of expanding

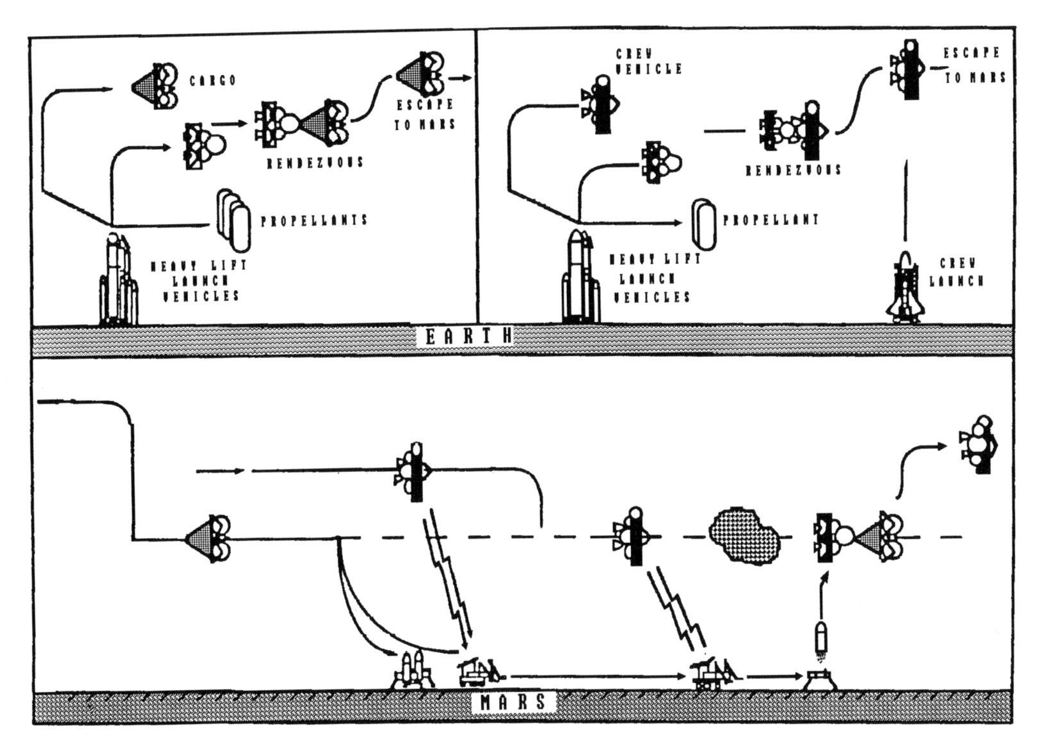

FIGURE 6.16: Schematic diagram of an initial mission to land on Phobos as outlined by NASA's Office of Space Exploration in one of the case studies to develop a strategy for the manned exploration of Mars. (NASA)

human presence and activity beyond Earth orbiting missions into the Solar System becomes a global effort.

The critical path in the Rockwell plan is very similar to NASA's Pathfinder Case Study 4, from a lunar outpost to early Mars expeditions. Starting with the Space Station *Freedom,* the plan continues through the development of the Moon and Mars into the next century with selection of candidates for interstellar exploration and migration to nearby star systems.

In 1989, the Office of Space Exploration published details of an in-depth case study for an evolutionary strategy to develop Mars. Initial missions to the Martian satellites slipped a couple of years from the earlier case studies. The first mission to these satellites was this time proposed for Earth departure in 2004 with the first exploration of the Martian surface in 2006. The missions are illustrated schematically in figure 6.17). Three crew members would leave Earth in May 2004 with the aim of conducting a preliminary survey of Phobos and Deimos similar to the earlier case study. Also the use of an advanced teleoperated robot roving over the Martian surface would be tested.

The rover on the surface coupled with surveillance from orbit would gather data so that a site for a permanent base on Mars could be selected.

The crew-carrying vehicle would be configured to simulate gravity by spinning about a tether as soon as it left the Earth. Arriving at Mars in April 2005, the crew stop the spinning and prepare the spacecraft for an aerobraking maneuver through the Martian atmosphere to capture the spacecraft into an elliptical orbit around Mars which would later be circularized. Two members of the crew would transfer to a small exploration

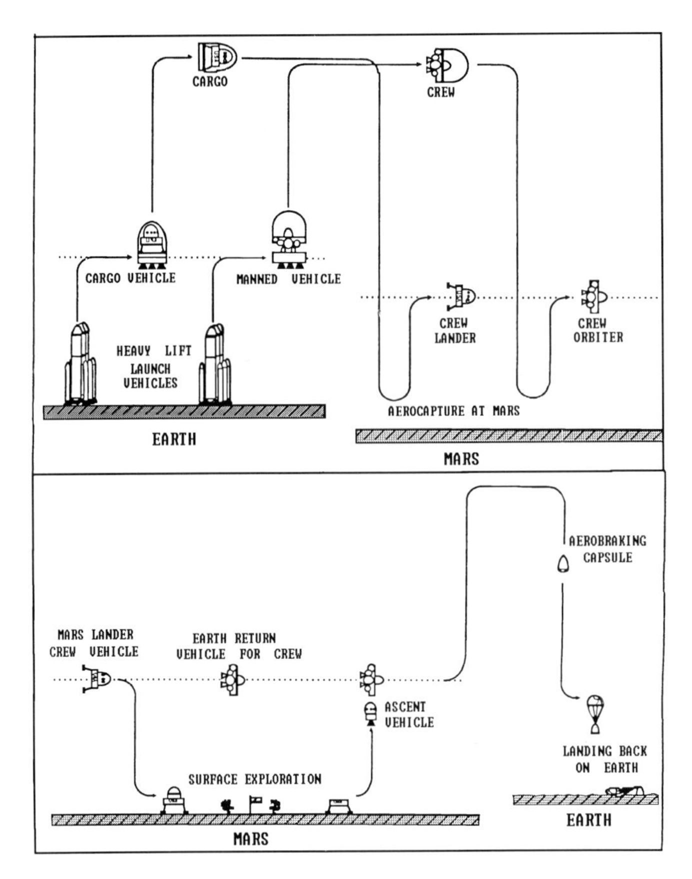

FIGURE 6.17: Schematic diagram of a representative mission to the surface of the planet outlined by NASA's Office of Space Exploration in one of the cases studied to develop a strategy for manned exploration of Mars. (NASA)

vehicle for descent to Phobos which would be explored and from which surface and core samples would be gathered. The Phobos mission would be aimed at selecting a site for production of propellants from the materials of the satellite.

Next the crew would use the satellite exploration vehicle to move to Deimos and conduct a similar exploration of that satellite with a similar objective. These two satellites are now regarded as the gateway to Mars (figure 6.18) as I first suggested more than thirty-five years ago in my 1952 book (see figure 4.4). A Phobos space station would provide a transportation node for activities on the surface of Mars. It would maintain a propellant storage depot and provide docking and hanger facilities for Mars ascent and descent vehicles and for interplatary spaceships.

During the exploration of the satellite, and afterward, the third crew member in the orbiting spacecraft would be controlling a large rover on the Martian surface. Samples collected by the rover would be transferred to an ascent module and brought back to the main orbiting manned spacecraft. The rover would be left on the surface for use by the next expedition which would land a crew on the surface of the planet.

After about three months in the Martian system, the crew would start back to Earth, arriving there in December 2005. The return vehicle would use aeroassist to enter Earth's atmosphere, slow down to LEO orbital velocity, and skip out of the atmosphere to rendezvous with the Space Station *Freedom*.

NASA's Project Pathfinder has suggested an autonomous lander to develop the

FIGURE 6.18: Gateway to Mars; the Phobos Space Station envisaged as a preliminary to the permanent human presence on Mars is here depicted by artist Mark Dowman of Eagle Engineering, Inc. The space station provides a transportation node for activities on the Martian surface. It maintains a propellant storage depot, and despun docking and hanger facilities. (NASA-Johnson)

technology required to ensure safe landing by aerobraking on Mars and safe return to orbit (figure 6.19). This would pave the way for placing a rover, or rovers, at interesting geological sites on Mars. The technology is also required to support autonomous resupply operations not only for Mars bases but also for bases on the Moon. Such a program, if approved, would focus on the development of terminal descent strategies and algorithms for the automated spacecraft's trajectories, as well as designs for guidance, navigation, and controls, their software and their sensing devices.

Following the success of the initial mission, a more advanced mission would carry five crew members to Mars using two main spacecraft, a crew carrier and a cargo spacecraft. The mission objective would be to land on the surface of Mars and explore the area selected as a site for a permanent manned base on the Martian surface. The flight profile is shown schematically in figure 6.20.

Departing from Earth in August 2005, the mission would arrive at Mars in February 2006. Note that the hardware for this mission must be fabricated along with that for the first mission, and a separate crew trained while the first crew is exploring the Martian system. At Mars the crew transfers to a landing and cargo vehicle carried by the main spacecraft and descends to the surface of Mars. The crew lives on the surface in a roving vehicle carried to Mars on the first expedition to explore the Martian satellites.

After its exploration of the site is completed and required samples have been gathered, the crew secures the rover and transfers to the ascent vehicle. Its stay on the surface could extend up to 200 days. If the crew decides to stay beyond that, it would have to wait another 300 days before it could return to Earth. The crew leaves the surface in the ascent vehicle and rendezvous in orbit with the main spacecraft, to which it transfers for the trip back home. As with the first expedition the returning spacecraft uses Earth's high atmosphere for braking and capture into LEO to transfer the crew and samples to the Space Station *Freedom.*

Exploration and development of Mars continues with a third and unmanned mission which leaves Earth in 2007 with a cargo destined to be placed on one of the Martian satellites to produce propellants there, and a habitat to be placed on Mars. It arrives at Mars in November 2008 and uses aeroassist to enter an orbit around the planet. The propellant production plant is landed on the most suitable of the two satellites, most

FIGURE 6.19: An autonomous lander has been studied by NASA as part of a Pathfinder Project to provide the technology required to ensure a safe landing at geologically interesting sites on Mars. Often such sites are hazardous. The program focuses on the development of terminal descent strategies and trajectory algorithms, as well as on guidance, navigation, and control designs, on software and on sensors. (NASA-Johnson)

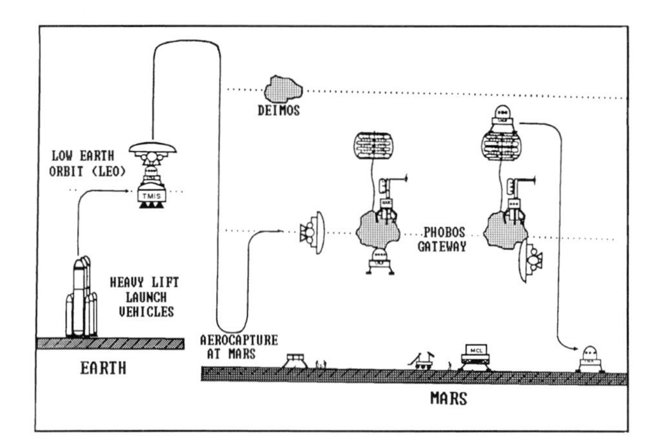

FIGURE 6.20: Flight profile for a major assault on Mars as developed during case studies by NASA's Office of Space Exploration. The objective of such a mission would be to land crews on Mars and select a site for a permanent manned base there. (NASA)

probably Phobos, and automatically begins operation to produce water and rocket propellants.

The fourth mission in the NASA scenario can next leave Earth. This takes place in October 2009 with the intent of setting up a permanently manned base on the surface of Mars. Its five-member crew arrives at Mars in July 2010. Propellants are obtained from the satellite propellant plant and the crew prepares for moving itself and cargo to the surface of the planet. A cargo lander descends first, followed by the crew-carrying lander. After a year on the surface, getting the habitat into order for human habitation, the crew returns to orbit in August 2011 and arrives back at Space Station Freedom in July 2012.

Meanwhile the fifth Mars mission has been getting ready. This duplicates many of the aspects of the fourth mission with the aim of having a human presence on Mars for a second year. A sixth mission is an unmanned cargo vehicle that leaves Earth around 2014 carrying major equipment to expand the facilities on Mars and on the Martian satellites. It is expected that this cargo vehicle will use nuclear electric propulsion, a technology that should be developed by that time. The cargo freighter will not only carry equipment to Mars but also bring back propellants from the Martian satellites to the Earth-Moon system, it being more economical to transfer rocket propellants from the Martian satellites to Earth orbit than from Earth's surface to Earth orbit.

The seventh mission carries a crew of seven to extend human operations on the surface. It will use nuclear electric propulsion to shorten considerably the transit time from Earth to Mars. It will leave Earth orbit in May 2016 and arrive at Mars just under four months later. The crew will be able to stay on Mars for nearly two years (640 days), arriving back in Earth orbit in September 2018.

Although this specific program has not yet been authorized as a national objective, a permanent human presence on Mars can be established in this way. A base on Mars is intended to ensure a continued human presence on the planet, rather than separate missions of the type exemplified in the *Apollo* program for preliminary exploration of selected localities on the Moon. A base is essential for any meaningful exploration of Mars either on a scientific basis or a long-term development basis. Initially the intent will be to obtain as much scientific information as possible about Mars—its present, past, and likely future. Eventually humans will undoubtedly want to develop Mars to tap its resources.

A Mars science workshop held at NASA-Ames Research Center in August 1989 developed a strategy for human exploration of the planet based on the science expected from such exploration. In a timetable suggested by the workshop, the human exploration would begin after the conclusion of precursor unmanned missions early in the twenty-first century, namely, about 2015. After a fifteen-year period of human exploration, full human utilization of Mars would commence about 2030. The workshop also suggested that a base for human operations possessing the most opportunities for scientific work is within the Coprates Quadrangle region which encompasses the Tharsis volcanoes, the great canyons, some ancient cratered terrain, and various outflow and runoff channels. Additionally, the area possesses what may be a 150-kilometer diameter paleolake within the highlands of the Margaritifer Sinus.

Water for life support and for other uses can be produced from the Martian regolith's permafrost or hydrous phases. The system would consist of a feed device, a thermal reactor, a fluidized bed, and a water collection device. This system should be capable of automatic, untended operation to produce about 10 tons per year for the life support system.

A controlled ecological life support system is needed in which food production becomes part of the life support system. It must be capable of controlling trace contaminants of chemicals and microbes, be regenerative and not consume expendable resources. The system must identify contaminants and issue warnings to the crew if these contaminants begin to be uncontrollable by the system. The system must also provide for the control and processing of biological and other waste material reducing them to useful products such as water, carbon dioxide, and nitrogen. Potable water and water for hygiene must be recovered as much as possible without relying upon chemicals and filters that would have to be supplied from Earth.

Landing sites for spacecraft must be established away from the habitats and must be provided with access roads and with navigational aids such as lights and radio beacons. Propellant storage and transfer facilities must also be established.

There are several requirements for power-generating systems; in the low range of less than 50 kw a photovoltaic system may suffice. This must be highly efficient, of small size and, of course, light in weight. Excess power could be stored in oxygen/hydrogen fuel cells. Large power demands could be met by a nuclear reactor that would yield up to 1 Mwatt. It would require radiators and provide continuous power through several Stirling-type engines. Radiation shielding would be obtained by burying the reactor in the Martian regolith. Mobile power would be needed for rovers, utility vehicles, and for mining operations. Some applications would use rechargeable systems such as batteries; others would use photovoltaic devices. All would have to be lightweight and highly reliable and capable of being serviced in the field.

Vehicles for use at the base include a pressurized rover, mentioned earlier in this chapter, capable of carrying a crew of two people over a 60-mile (100-km) round trip in a few days. The addition of trailers could extend the range of operations and their duration. Such trailers might include a habitation trailer and an additional power-generating trailer. This rover should be capable of carrying an additional two crew members if needed for a rescue operation. Airlocks should allow crew members to operate outside the vehicle also. An unpressurized rover, similar to the lunar rovers, would allow suited crew members to move around the environs of the base for operations on the Martian surface in support of the base or experiments to be performed on the surface near the base. It would have a range of about six miles (10 km) and be capable of carrying up to four people and a variety of payloads. A truck will be needed to transport cargo over distances of up to three miles (5 km). It would be used, for example, to move regolith to shield the habitat. It could also be used to transport the habitat modules from the landing site to the base site.

Christopher P. McKay of NASA-Ames Research Center, and Thomas R. Meyer of Boulder Center for Science and Policy, have made a number of continuing studies concerning the use of Martian resources for human settlement on the planet. Self-sufficiency of the Martian base is an important goal since this will lower costs of operations on Mars and the maintenance of humans and their machines and instruments on the planet. In addition, a base will provide better security and safety than individual missions. To make a base self-sufficient will require extensive use of Martian resources, the production on Mars of bulk consumables, propellants and fuels, potable water, oxygen, food, building materials, fertilizers, chemicals, and energy.

Power on Mars is a limiting factor in terms of joules per unit mass and the energy needed to make resources from Mars materials. As much as 10kw of power may be required to process each kilogram of Martian material.

Nuclear power plants are not a panacea; they are very massive because of their shielding. But reactors could be brought directly from Earth in cargo vessels and landed in advance close to the site selected for the base. Small reactors weighing about 16,500 pounds (7,500 kg) can produce 300 kwatts for ten years or so with very little maintenance. This may be the initial way of producing power on Mars, although solar power and wind power may be preferred later.

Solar cells would be seriously affected by the dust storms. Also they are bulky, must be provided in collectors of large area, and have a limited lifetime. However, a considerable amount of light does penetrate the dust storms for conversion by solar cells. Solar reflectors might be used to concentrate solar energy in closed cycle turbine engines. However, it must be remembered that the solar constant at Mars varies considerably as Mars moves from aphelion to perihelion; between about 500 and 700 watts per square meter at the top of the atmosphere.

Wind power is another possibility if the site for the wind-driven generators is carefully selected. Some sites may have winds of 20 miles per hour (32 km/hr), which would be sufficient to produce about 15 kwatts for each 1076 square feet (100 square meters) of windmill area.

Another power-related problem is that there is no oxygen in the Martian atmosphere and no fossil fuels in the surface rocks. Fires cannot be lit outside on Mars, nor can internal combustion engines use the Martian air as an oxygen supply. Its low pressure would also make supercharging necessary to take in sufficient mass for a working fluid to be expanded by heat from an oxidizer and fuel stored in a vehicle's propellant tanks.

Electricity provided by the power system could be used to split atmospheric carbon dioxide into oxygen and carbon monoxide, which could provide propellants for vehicles and to refuel rockets for return to Earth. Off-the-shelf equipment is available for this purpose. Electricity can also be used to split water into hydrogen and oxygen for use as propellants. Water will, however, initially be very precious as water, and it may not be acceptable to dissociate water to produce oxygen and hydrogen although these would provide a higher energy propellant combination. To produce propellants from the atmosphere sufficient for a returning crew to rendezvous with an interplanetary spacecraft, many tons of atmosphere would have to be processed assuming a reasonable efficiency of operation.

Somewhat more ambitious than the NASA scenario for manned missions to Mars, an initial crew of 15 was proposed in the Case for Mars studies discussed earlier. This crew would have to remain on Mars for two years and this would require that the first crew would need to bring supplies to last at least for that period. Should it be found impractical to replenish these supplies on Mars, subsequent missions would also have to use cargo ships to bring in bulk consumables from Earth. This would considerably increase the cost of maintaining a permanent base on the Red Planet.

This first crew has the arduous task of building the base from scratch, using the materials brought from Earth together with Martian materials such as soil to protect the habitats from radiation. It will have the subsequent task of making the base operational, setting up the power-generating equipment, building greenhouses and planting seeds, installing the equipment to extract carbon dioxide, oxygen, and water from the Martian atmosphere, and setting up the communication links with Earth.

The shells of the cargo vehicles will be designed so that they can become habitats. Furnishings will be installed in the habitat sections before the cargo ships leave Earth orbit. Each cargo vessel will provide a habitat sufficient to house five crew members with their personal quarters, general living areas, and kitchen facilities. Other cargo shells will provide laboratory space and possibly recreational facilities including exercise equipment and a comfortable library. The cargo vehicles, once they have been unloaded, will be dragged on skids to where they are to be located in the base complex. They are fitted with airlocks and interconnected by tunnels. Items freighted from Earth are placed in them, power sources set up and turned on. Equipment to extract resources are set up and activated. Next, draglines and buckets will be used to move Martian soil to cover the habitats and protect the crew from solar protons and cosmic rays. Airlocks provide access to the Martian surface and to air-shell structures. The life support system is set up and activated to provide air and water recycling, with reservoirs and the ability to make up for leakages. Vital systems must all have backup redundancy to guard against catastrophic failures. Much of this technology will have been developed and tested in operating the *Freedom* Space Station.

As in the NASA scenario, the Case for Mars base will be supported by several vehicles brought to Mars in the cargo ships; rovers for scientific exploration, truck/tractors to move equipment, and a base science vehicle containing science equipment that is too large or too heavy for use in the smaller rovers which are intended to have a much greater range of operations over the surface.

Breathable air will be derived from the atmosphere by reducing the carbon dioxide to obtain oxygen from it. As on Earth nitrogen can be obtained directly by cooling the atmosphere and fractional distillation. The remaining carbon monoxide can be used as a propellant.

Extraction of gases from the atmosphere can operate unattended at virtually any site on the planet, whereas mining from the soil needs careful site selection. Indeed, atmospheric "mining" might be started before the arrival of the first humans. Atmospheric processing equipment could be landed in advance of the manned mission and start to produce and store supplies of consumables under automatic control.

Water can be obtained directly from the humid atmosphere, but this requires adequate energy to process the large volumes of atmospheric gas. Although the atmosphere of Mars is often saturated with water vapor, it contains only an extremely small amount of water. In fact, if all the water in the atmosphere on the average were condensed into a liquid on the surface it would provide a layer of water only about 50 micrometers deep. At lower elevations there is more water in the Martian atmosphere, up to twice as much as average in the deep Hellas Basin, which averages about 2.5 miles (4 km) below the mean surface level of the planet. While atmospheric water may be sufficient to meet the needs of the initial base, especially if that base practices rigorous recycling, a permanent human presence on the planet will require greater supplies of water. Consequently the base will probably tap the permafrost for its water supplies as more water is needed. However, the evidence of a permafrost rich in water applies to latitudes greater than 40 degrees. The equatorial regions may not have this water supply, which should be taken into account when planning the location for a permanent base. It is also possible that liquid water reservoirs exist below the permafrost at a depth of several thousand feet, and these reservoirs may ultimately be pumped directly to the surface by sinking deep wells. There are also possibilities of brine solutions existing fairly close to the surface. These, too, might be accessed by wells and the brines desalinified so that the water can be used by the base.

An essential piece of equipment to process the atmosphere is a compressor that will also serve to pressurize the habitats. Backup will, of course, be necessary. Waste processing facilities and noisy equipment will be located separate from the living and working habitats.

The Case for Mars studies suggest the use also of lightweight air shells, erectable structures that are held rigid by internal pressure of compressed Martian air. Within such a shell personnel could work at normal pressure with an oxygen supply no bulkier than those used in scuba diving. These structures would provide greenhouse space, space to repair rovers and movable equipment, and semiprotected space for scientists making some scientific experiments. Protection from ultraviolet would be required and could be achieved by suitable choice of material for the air shells. This material would also have to be selected on the basis of its durability under Martian conditions; it must not degrade under solar ultraviolet radiation or from Martian dust abrasion.

Food can be produced initially within the greenhouse air shell structures (figure 6.21), using filtered sunlight or artificial light. The soil of Mars appears to have all the

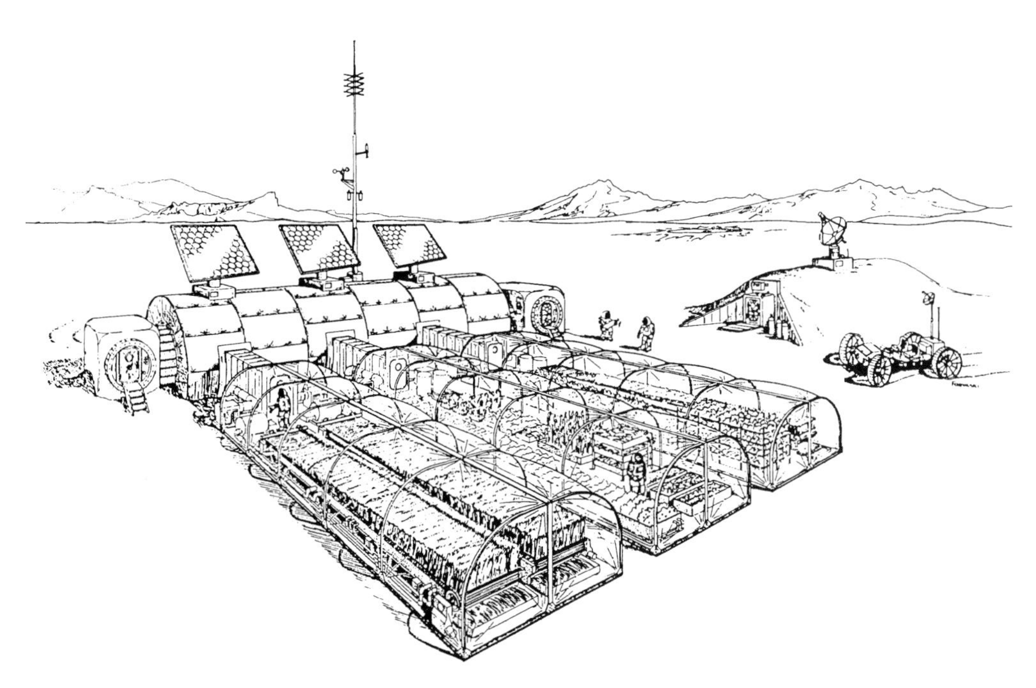

FIGURE 6.21: Diagram of a Mars base showing air shells acting as greenhouses in the foreground and a habitat covered by Martian soil in the background. A simple type of rover vehicle for operating close to the base, as opposed to distant exploration, is also shown. Airlocks provide access to the Martian surface for the spacesuited crew members. (Carol Stoker, NASA-Ames)

necessary elements needed for plant growth. Heating would be derived from waste heat from the power generating system, the compressors, or the habitats. Later food production from plants may be moved completely underground to obtain better environmental control.

Food, water, and a breathable atmosphere are three fundamental requirements for humans to survive. Expendable supplies of these elements can be carried on board for brief space flights. Space stations in low Earth orbit can be supplied from the Earth at regular intervals. But for interplanetary journeys and a base on Mars, food, air, and water must be regenerated. A controlled ecological life support system (CELSS) is being developed by NASA for this purpose. Nearly 90 percent of human diet for energy consists of carbohydrates and fat. Edible fats can be synthesized through chemical processes, but techniques for synthesizing carbohydrates are only just beginning to be investigated. Crop plants and other photosynthetic organisms produce large amounts of carbohydrates.

Central to the CELSS concept is a plant growth unit in which a crop of wheat has already been brought to harvest successfully. Plant biomass produced in a plant growth unit must be passed to a food-processing unit where the edible materials are converted into prepared foods. When these food products are consumed by the crew and metabolized into carbon dioxide, water, urine, and fecal material, all must be recycled. Solid and liquid wastes are transferred to a waste-processing system and converted into chemical forms that are usable as plant nutrients. Carbon dioxide and water vapor are recovered from waste solids by oxidation. Water is cycled between the plant growth unit, food-processing system, crew, and waste-processing system through a dehumidifier and a purifier.

Trying to recreate the cycles of nature in a small environment such as a Martian base is undoubtedly a great technical challenge. Nevertheless, great progress has already been made. Wheat has been grown at NASA-Ames Research Center in a simulated Martian atmosphere supplemented with oxygen. Techniques have also been developed to recover pure protein from algae grown under controlled conditions similar to a Martian base. Experiments seem to show that food crops can be grown on the native soil of Mars if it is supplemented with nutrients and some minerals and its oxidant is neutralized by water. Research with CELSS has already shown that the concept of providing a self-contained Martian life support system is a valid concept. The crews at the base should be able to "live off the land" and conserve their precious water and oxygen by efficient recycling.

When the initial setup of the base has been completed, the crew commences its long-term activities; observing and forecasting Martian weather, monitoring reactions of the crew members to life on Mars, checking the territory around the base for permafrost, minerals, and other natural resources to be used in expanding the base. At the same time crew members will start their scientific work, collecting and analyzing samples, setting up stations to record seismic and meteorological conditions, searching for evidence of past or present microbial life or even for more advanced life forms, planning and executing long-range roving missions over the surface. They will report to Earth on the progress of the base and of the exploration and will make recommendations about the mission being planned for the next Mars opportunity.

How soon can this exploration of Mars begin? Whenever the nation decides to start. There is no lack of visionaries and optimists for a peaceful and fulfilling future as humans expand into and explore the Solar System. Orbiters and rovers can be sent to Mars in the 1990s to make the necessary preliminary surveys. A sample of Mars could be returned to Earth within ten years. The first manned mission could be sent to a Martian satellite, Deimos or Phobos, by the year 2005, and a manned mission to the Martian surface could take place before the year 2010. Subsequently crews could be rotated by missions every two years. Humans would be a new and permanent life on Mars for as long as they wished; ultimately, perhaps, to become independent of support from their home planet, Earth. The Martians would at last have arrived! From Mars the human experience can be expanded into the Solar System and beyond. The species has the opportunity of developing into an interplanetary life form with all that that implies as it reaches toward the beckoning stars.

7

PROMISE OF THE FUTURE

There is a growing awareness among many young people of today's world that space is the new econiche for mankind, that space is where the action of the future is going to be. Expansion into space is seen as the great opportunity of our age, comparable to that Renaissance period in human history when people began to expand from the European continent to develop all of this planet. Indeed, it is often stated that space can provide a safety valve, just as the Americas and other continents served our forebears when they were frustrated by European squabbles and hidebound traditions. Now we have before us the challenge presented by a whole universe waiting to be explored and developed, an unlimited Lebensraum to relieve the social and political pressures of a planet primarily engaged in defending itself against itself, like a biological creature with its immune system running amok.

Unfortunately there is also much negative thought about space and our future: the so-called cost effectiveness of space exploration, the dollar value of any expansion into the Solar System, claims that we should clean up the sewers before reaching for the stars. An analogous situation prevailed in Europe as hardy pioneers began to reach out for the other continents of Earth. Conditions were atrocious: stinking sewers, rampant sickness and poverty, interminable wars, social injustices, and economic disasters—indeed, every justification for staying at home and ignoring the new opportunities overseas. Fortunately, the spirit of exploration prevailed. Opportunities were seized by some nations and neglected by others. New worlds of Earth were investigated and developed. Broadened horizons reflected back to Europe and caused great social changes to trigger a subsequent era of prosperity.

The Space Shuttle and the big unmanned boosters were the beginning. Other more advanced space transportation systems will open a new era in the evolution of our species during which travel to new worlds of space can become as commonplace, if we wish, as jet flight between continents is today. Space transportation systems also provide an opportunity for people to build in space as in the past they built on the Earth. We can construct space hotels and recreation centers, gravity-free havens for paraplegics, cities on other worlds, artificial planets, starships. While the builders of henges and pyramids dragged stones across the Earth, the builders in space will more easily move materials about the gravitational plains of the Solar System, mining the asteroids if needs be.

Building such structures is a long-term undertaking, comparable in terms of gross national product to building Stonehenge, the Gothic cathedrals, Mayan ceremonial centers, and the Panama Canal. It demands the same kind of national and human dedication. But while large projects in space may sometimes seem beyond our political and social systems to accomplish, they are not beyond our technical skills. The people of this nation worked modern miracles of technology and management during a decade of concerted effort that enabled men to walk on the Moon.

The *Apollo* program revitalized the U.S. economy, as mentioned in an earlier chapter. The manned mission to Mars could be accomplished today, building on the technology, science, and experience of *Apollo* at a cost of only one-quarter of the cost of the *Apollo* program in terms of gross national product. For an investment of a mere half percent of the current U.S. gross national product this nation could go down in history as the one that opened the Solar System's vast resources to humanity. What a contrast this accomplishment would be to being the nation that first puts "stars wars" technology and nuclear weapons into space! The choice is ours.

Today we have succumbed to a terrible mental miasma; if something can't be done right now, or if it changes our environment, it is not worth doing unless it enhances "national security." But this is not the story of life. All living things change the environment as they reverse growth of entropy in their local biological system. Most living things also dedicate their lives to the future, and seek to preserve their species by gradually changing their progeny so it can adapt to the ever-changing environment of a dynamic planet such as Earth. Otherwise they would become fossils, in common with the majority of other life forms that have populated this planet since the origin of life here.

Today we often seem to lack the optimism of our ancestors. We are like acorns scared of developing into oak trees because we might be chopped down. We tend to forget the story of our species, the examples of dedicated people who were willing to use a whole lifetime if necessary to try to achieve a clearly defined purpose—to paint the ceiling of the Sistine Chapel, to build transcontinental railroads, to develop airplanes, to raise a new generation of human beings to a more fulfilling maturity. With optimism we can stand on the shoulders of those giants and achieve even more today.

We can reach for the stars—physically, intellectually, and spiritually. But we need a national plan and bold leadership. We need new goals, not only for the nation but also for humankind—goals that will stretch out to the very limits of the universe and into the depths of the atom and the living cell.

Critics say we have insufficient money to do these things. They forget we invented money. It is not money we need, for governments and individuals have traditionally always debased the symbolic value of currencies. The need is for wealth, and fortunately we have unlimited wealth. It is of two kinds—physical and metaphysical. Physical wealth is the ability to utilize energy in the manipulation of matter, forces, and fields. There is an almost unlimited amount of energy in our universe challenging us to tap it at the center of atoms and the centers of galaxies. Aggressively seeking even a few extraterrestrial energy sources will break us from our negative thoughts of limits to growth and human fullfillment.

Metaphysical wealth consists of the ideas derived from an understanding of what makes the universe tick, the principles that govern galaxies and stars, subatomic particles and molecules, and the microbiology of living things. These principles have always been around waiting to be understood. They were available to consciousness from the beginning of time. But we must continually probe deeply and extensively with our minds through both space and time if we are to find and understand them.

It is important to accept that people have a vital role to perform in space. We cannot expand into space with automated machines alone, unless we abdicate in favor of von Neuman machines able to replicate themselves and spread a machine-based culture throughout the galaxy in place of a bio-culture such as ours. We are one of the few terrestrial creatures having the ability to look at the stars. People have always looked at the stars—perhaps because they instinctively know that their destiny is out there. It could very well be that humanity has a future in space, perhaps an important role to play in making the universe aware of itself. Beyond the planets we can even reach for the stars, touching them physically as in the past we have touched them with our minds through studies of astro and nuclear physics.

It may seem that current technology is impractical for interstellar missions, even with unmanned spacecraft. However, it would be unwise to declare that flight to nearby stars is impossible. In the late fifties many reputable engineers and scientists "proved" that those ballistic missiles which now threaten sanity on this planet were unlikely to attain intercontinental ranges, and that probes to other planets and manned flights to the Moon had to be classed as fantasy. This was a decade after Arthur C. Clarke and the author had won first and second prizes in a *Royal Air Force Quarterly* essay competition on the future development of such long-range missiles, and had published the first definitive papers on manned and unmanned communications satellites respectively.

If the space age could be missed by so-called experts only four years before the Soviet *Sputnik I,* the first man-made satellite, streaked through the constellations in October 1957, it would seem unwise today to dismiss all ideas of interstellar missions as unrealistic dreams.

But first we have to expand into the Solar System, establish space settlements, build bases on other worlds, on the Moon and on Mars, and if necessary, change the environment of some planets so that interplanetary humans can thrive on them and we can gain experience as being extraterrestrials (figure 7.1).

If we as a species survive over the next twenty-five years or so without a major catastrophe, nuclear or otherwise, natural or man-made, we shall begin to reach for the stars, first with settlements throughout the Solar System, and then with unmanned probes to search for habitable planets of other star systems. Later we will build starships for manned expeditions to and the colonization of other planetary systems.

There are several ways in which this might be done, all of which have figured in science fiction stories for decades. We could make artificial planets that require centu-

FIGURE 7.1: If we as a species can survive over the next 25 years or so without a major catastrophe, nuclear or otherwise, natural or man-made, we shall begin to reach for the stars and gain experience on other worlds with broadening horizons of a whole universe out there waiting to be explored and possibly to be understood. (Artwork courtesy NASA-Johnson)

ries of travel to reach our destination so that only the descendents of the original space travelers become colonists.

We might make smaller starships in which the space travelers are placed in suspended animation for the journey. There is also the possibility of developing advanced biological systems to prevent aging so that the star travelers would be volunteers willing to spend centuries of their immortality voyaging to other star systems.

We may be able to design high technology starships that travel at a sizable fraction of the speed of light. Perhaps people can be transmitted to the stars as encoded information. We can even imagine a tachyon transfer in which people jump about the universe instantaneously for reassembly wherever they wish.

If we build on our space experience and settle our ridiculous political and territorial squabbles here on Earth, our future in space can develop from the Space Shuttle to settlements in space, from settlements to extraterrestrial bases and colonies on other worlds of the Solar System, and from the Solar System to interstellar exploration and migration—a veritable human seedcast into the relatively limitless territory of the Galaxy. It is vital that we encourage young people of the space age to become involved in the excitement and the challenge of living today. They need assurance that they, too, can live their creative dreams as the rocket pioneers of the 1930s, lived their dreams and squeezed humankind out of the restrictive womb of Earth. We have a duty to turn the children of the space age toward the real experience of reaching for the stars as a viable and rewarding alternative to seeking a chemical fantasyland that pollutes their inner environments.

We are at a new beginning. There are, beyond Earth, unlimited economic commons awaiting our development in the new econiche of space. The lesson we should learn from biology is that when a new econiche is opened some form of life will jump into it. Humans have a purpose in space. There is an inherent biological force driving us—the life force, which forever seeks new and less competitive econiches into which to expand and flourish. While some people may resist, and many will choose to remain on Earth as many remained in Europe, as a species we may not be able to escape, except by self-destruction, what may be our programmed destiny of spreading consciousness throughout the universe.

This nation needs a new and challenging positive national goal, a bold long-range plan to take the spirit of 1776 and create a new era of human development, and unite the diversified people which are the strength of the nation. We need to reach for the beckoning stars with no limitations on our resources or on what we can accomplish. We need to abandon our current national preoccupation with material personal gains, to abandon emphasizing racial or cultural differences and ethnic roots, and to concentrate instead on a united human effort to reach for the stars. We need to encourage new and stimulating enterprises capable of demonstrating government leadership and returning people's confidence in the ability of our elected representatives to lead us into a future of optimism and well being for humankind. The permanent base on Mars is one of such enterprises. Through it we have the opportunity to lead humankind into the Solar System.

Make no mistake, however, there is competition for the new econiche. Other humans are anxious to carry their way of life to the stars. The Soviets have solved many of the technical and human problems associated with establishing a base on another world. The Chinese have an ambitious space program as well as a nuclear ballistic missile capability. The European community is encouraging much new space research including winged orbital transportation systems such as Hotol and Sanger.

The stakes are very high. On one extreme we have the ideology that the governing body (state or international conglomerate) is supreme and the individual is its absolute vassal. At the other we have the spirit of '76, where the governing body is the servant of the people and the rights of the individual are paramount. How do we want the Solar System, or the Galaxy, to be governed? As a super state or super corporation of ant-like workers? or as a society of unique creative individuals working toward a common good?

We are the result of a fascinating evolutionary process—atomic evolution within stars to complex atoms able to exchange energy in chemical reactions; molecular evolution on planets, the ashes of stars, to complex molecules that discovered how to replicate themselves; and biological evolution to organisms having the ability to question their evolution and their purpose.

Now we have entered a most important fourth phase of evolution, that of consciousness, in which we may have the opportunity to exchange thoughts with extraterrestrials. Certainly during this evolutionary period each of us should develop a capacity for creative fulfillment by which we refresh mankind with new goals and new values. The future is ours to fashion in any way we please—to reach for the stars or to descend into a new barbarism of greedy materialistic individuals working in paper-shuffling towers, and even greedier states and international corporations dictating how the papers must be shuffled.

Through space exploration following a well-conceived and bold plan, we can create new worlds and new opportunities for humankind, and new creative challenges by which individuals can achieve fulfillment instead of suffering frustration.

As we reach for our individual stars we can become more aware of creation and of that which creates, more aware of ourselves and of our opportunity to make a universe become aware of itself. In so doing we can escape the inevitable fossildom of a human species confined to Earth.

The universe is not friendly to us; nor is it unfriendly. It is impartial like a good referee. But we have to know the rules, the basic physical rules by which the game of life must be played within it. Physical exploration coupled with science—the mental interpretation of our observations—is essential to our discovering these rules of life on a cosmic scale by which we can enrich our lives.

A myriad stars beckon in the microcosm of the atom and the macrocosm of the galaxies and at all levels in between. As a nation we should boldly go forth to lead other peoples of the planet into the future of new opportunities beyond the limitations of planet Earth as we paraphrase those encouraging words uttered at another time and place: Be not afraid. . . . in my house there are many mansions. Perhaps Mars is the first of these mansions, providing a new viewpoint of man's place in the universe as we return life to the Red Planet.

SUGGESTED FURTHER READING

(Arranged chronologically)

1895 *Mars*, by Percival Lowell. Boston, Houghton Mifflin Co.
1906 *Mars and Its Canals*, by Percival Lowell. New York: The MacMillan Co.
1907 *Is Mars Habitable?* by Alfred Russel Wallace. London: Macmillan and Co., Ltd.
1930 *La Planete Mars*, by E. M. Antoniadi. Paris: Hermann.
1951 *The Sands of Mars*, by Arthur C. Clarke. London: Sidgwick and Jackson.
1952 *Rocket Propulsion*, by Eric Burgess. London: Chapman & Hall Ltd.
1954 *The Green and Red Planet*, by Hubertus Strughold. London: Sidgwick and Jackson.
1954 *Physics of the Planet Mars*, by G. de Vaucouleurs. London: Faber and Faber.
1954 *Exploring Mars*, by Robert S. Richardson. New York: McGraw-Hill.
1955 *Frontier to Space*, by Eric Burgess. London: Chapman & Hall Ltd.
1956 *The Exploration of Mars*, by Willy Ley and Wernher von Braun. London: Sidgwick and Jackson.
1956 *A Space Traveler's Guide to Mars*, by I. M. Levitt. New York: Henry Holt and Company.
1962 *Mars*, by Earl C. Slipher. Cambridge: Sky Publishing Corporation.
1965 *Mars*, by Robert S. Richardson and Chesley Bonestell. London: George Allen & Unwin Ltd.
1966 *Mariner IV to Mars*, by Willy Ley. New York: Signet Science Library.
1967 *A Review of the Mariner IV Results*, by Oran W. Nicks. NASA SP-130, U.S. Government Printing Office.
1967 *Mariner-Mars 1964 Final Project Report*, by Jet Propulsion Laboratory. NASA SP-139, U.S. Government Printing Office.

1968 *The Book of Mars,* by Samuel Glasstone. NASA SP-179, U.S. Government Printing Office.

1969 *Mariner-Mars 1969,* by Jet Propulsion Laboratory. NASA SP-225, U.S. Government Printing Office.

1971 *The Mariner 6 and 7 Pictures of Mars,* by Stewart A. Collins. NASA SP-263, U.S. Government Printing Office, 1971

1973 *Is There Life on Mars?* by Graham Berry. Los Angeles: The Ward Ritchie Press.

1974 *The New Mars,* by William K. Hartmann and Odell Raper. NASA SP-337, U.S. Government Printing Office, 1974

1974 *Mars as Viewed by Mariner 9,* by Principal Investigators. NASA SP-329, U.S. Government Printing Office.

1976 *A La Recherche d'une vie sur Mars,* by Albert Ducrocq. Paris: Flammarion.

1976 *On the Habitability of Mars,* by M. M. Averner and R. D. MacElroy. NASA SP-414, U.S. Government Printing Office.

1976 *The Search for Life on Mars,* by Henry S. F. Cooper. New York: Holt Rinehart and Winston.

1977 *Mars at Last!* by Mark Washburn. New York: G. P. Putnam's Sons.

1977 Scientific Results of the Viking Orbiter, *Journal of Geophysical Research,* Vol. 82, No. 28, September 30, 1977, reprinted as a book, American Geophysical Union, 1977

1977 *The Martian Landscape,* by Lander Imaging Team. NASA SP-425, U.S. Government Printing Office.

1978 *To The Red Planet,* by Eric Burgess. New York: Columbia University Press.

1978 *Mars!* by Jeff Rovin. Corwin Books.

1979 *Life on Mars,* by David L. Chandler. New York: E.P.Dutton.

1979 Proceedings of Second Mars Colloquium, Jan. 1979, *Journal of Geophysical Research,* Vol. 84, No. B14, Dec. 1979, American Geophysical Union, 1979

1979 *The Atlas of Mars,* by U.S.G.S., NASA SP-438, U.S. Government Printing Office, 1979

1979 *Handbook of Soviet Lunar and Planetary Exploration,* Nicholas L. Johnson, Ed., American Astronautical Society. San Diego: Univelt, Inc.

1980 *Volcanic Features of Hawaii,* A Basis for Comparison with Mars, by M. H. Carrand and R. Greeley. NASA SP-403, U.S. Government Printing Office.

1980 *Viking Orbiter Views of Mars,* Cary R. Spitzer, Ed. NASA SP-441, U.S. Government Printing Office.

1980 *Images of Mars,* by Michael H. Carr and Nancy Evans. NASA SP-444, U.S. Government Printing Office.

1981 *The Surface of Mars,* by Michael H. Carr. New Haven: Yale University Press.

1982 *Mission to Mars,* by James Oberg. New York: New American Library.

1982 *The Channels of Mars,* by Victor R. Baker. Austin: University of Texas Press.

1982 *The Case for Mars,* Penelope J.Boston, Ed., American Astronautical Society. San Diego: Univelt, Inc.

1983 *Earth's Earliest Biosphere; Its Origin and Evolution,* by J.W. Schopf, Ed., Princeton University Press.

1985 *The Case for Mars II,* Christopher P. McKay Ed., American Astronautical Society. San Diego: Univelt, Inc.

1986 *Pioneering the Space Frontier,* by National Commission on Space. New York: Bantam Books.

1986 *Planetary Exploration Through the Year 2000,* Part II, by NASA Advisory Council. U.S. Government Printing Office.

1987 *The Monuments of Mars,* by Richard C. Hoagland. Berkeley: North Atlantic Books.

1988 *Race to Mars,* by Frank Miles and Nicholas Booth. New York: Harper and Row.

GLOSSARY

AEROASSIST: Use of the high atmosphere of a planet to maneuver a spacecraft by changing its path or its velocity or both without the spacecraft necessarily plunging deep into the atmosphere.
AEROBRAKING: Use of aeroassist to slow the motion of a spacecraft especially for changing an orbit around a planet or for a descent into deeper regions of the atmosphere or to the surface of a planet.
AEROCAPTURE: Use of aeroassist to change the path of a spacecraft into an orbit around a planet.
ALBEDO: A measure of the ability of the surface of a planet to reflect radiation incident upon it. A perfect reflector has an albedo of 1.0 and perfect absorber has an albedo of 0. In visual light, an albedo of 1 is a pure white and an albedo of 0 is an absolute black.
ALLUVIAL DEPOSIT: Sediments which were transported by a flow of water and deposited as the flow could no longer carry them further.
AMINO ACID: Building block for large organic molecules such as proteins. All amino acids in terrestrial life forms are of one type. Some amino acids discovered in meteorites are of a different type, like a left hand glove compared with a right hand glove.
ANAEROBIC BACTERIA: A microbial life form which does not require oxygen for its metabolism.
ANHYDROBIOSIS: The ability of a living organism to survive in a dehydrated (waterless) state for a period and then return to a normal living state when again provided with water.
ANTHROPOMORPHIC: Application of human attributes to nonhuman things and situations and to pantheons of deities.

APHELION: The point in the orbit of a body moving in a noncircular path around the Sun which is most distant from the Sun.
APOAPSIS: The point on the orbit of a body which is moving in a noncircular path around another body, such as a planet or a satellite, which is most distant from that body.

BAR: A unit of pressure; Earth's mean atmospheric pressure at sea level is 1.01325 bar.
BASALT: A dark-colored, fine-grained volcanic rock. On Earth it forms the bedrock of all of the oceanic crust and it is also present as extensive flood sheets on the continents.
BASE SURGE CRATER: An unusual crater form which is found on Mars and has not been seen elsewhere in the Solar System in which a fluidized apron seems to have extended around the crater. This is generally assumed to have been caused by the presence of volatiles, generally water, in the surface materials into which the impacting body dug the crater.
BIOGENIC COMPOUND: A molecule that is basic to being incorporated into the structure of living things.
BIOSPHERE: The region of a planet that can support life and includes living things. On Earth this includes the atmosphere, the oceans, and the surfaces of the continents extending partway into the soil. The terrestrial biosphere is an extremely thin shell around the planet in relation to the size of the planet.

CALDERA: A large, roughly circular volcanic crater with steep sides and possessing a flat floor formed by cooled lava flows.
CARBONACEOUS CHONDRITE: A type of meteorite rich in carbon compounds.
CARBONACEOUS MATERIAL: A material rich in carbon or carbon compounds.
CATALYST: A substance which causes a chemical reaction without entering into or being changed by that reaction.
CHAOTIC TERRAIN: A jumbled terrain seen on Mars which appears to have resulted from the effects of volcanic heating removing water from ice-rich subsurface material.
CRATERED TERRAIN: Ancient surface of a planet which still shows many craters believed to have resulted from bombardment of the planets of the Solar System and their satellites soon after their formation several billion years ago.

DENDRITIC CHANNEL: A valley stream with tributaries merging into a main channel which in plan form resembles the branching pattern of some trees.
DIFFERENTIATION: The gravitational separation under heating of the material of a planet into shells with lighter materials rising toward the surface and heavier materials sinking toward the core.
DNA: Deoxyribonucleic acid, a large molecule consisting of patterns of four amino acids which carries the genetic code to determine the specific form of a living organism.

ENZYME: A protein manufactured by living cells which acts as a catalyst for chemical reactions within living organisms while not changing itself in the process.
EOLIAN: Pertaining to the effects of wind, usually in the sculpting or molding of surface features.
ESCAPE VELOCITY: The velocity at a position in the gravitation field of a planet which a body must attain for it to travel to an infinite distance from the planet without further input of energy. Usually applied to the velocity required at the surface of a planet or other body for a projectile to escape from the planet or other body. The escape velocity at the surface of the Earth is, for example, 6.93 miles per second or 11.18 km/sec. That from the surface of Mars is 3.1 miles per second or 5.0 km/sec.
EXOBIOLOGY: The study of life beyond Earth, its possibilities, its likely development, and the search for such life.

FLUVIAL ACTIVITIES: The action of water flowing in channels such as streams and rivers.

GAS CHROMATOGRAPH MASS SPECTROMETER: An instrument to separate different compounds carried by a flow of gas and then analyze these compounds by their masses.
GAS EXCHANGE EXPERIMENT: Soil samples placed in a test chamber were incubated with water vapor and then wetted with a nutrient solution to check if gases in the space surrounding the sample were changed in composition by biological activity in the soil sample.
GREENHOUSE EFFECT: Trapping of solar radiation within the atmosphere of a planetary body by gases which permit radiation to enter the atmosphere but prevent the escape of heat through the atmosphere, analogous to the heating effect within a greenhouse.

HADLEY CELL: Rotation of atmospheric gases from equator to pole at high altitude and then back to the equator at low altitude.
HYDRATED MINERAL: A mineral containing water in which the water is chemically combined with the rock materials.

IMPACT BASIN: Large circular flat-floored depression surrounded by concentric mountain ranges, formed by the impact of an asteroid-sized body on a planet or a large satellite. Typical impact basins are the Mare Imbrium on the Moon and Hellas on Mars.
IONOSPHERE: Region of ionized gases in the low density upper atmosphere of a planet resulting from incoming solar radiation ionizing atmospheric molecules and atoms to produce free electrons and positively charged ions.
ISOTOPE: Form of an element which although having the same chemical properties differs in the contents of the atomic nucleus so that the atomic mass differs because of the presence or absence of some neutrons.

KELVIN: Degree of absolute temperature.

LABELED RELEASE EXPERIMENT: An experiment to look for microbial activity in a soil sample. It checked for metabolizing of labeled carbon-14 in a liquid nutrient and its release as carbon monoxide or carbon dioxide by such microbial activity.
LAYERED DEPOSIT: Materials deposited in layers by various weathering processes such as sedimentation in bodies of water or fallout from dust storms.
LIPID: An organic compound that is insoluble in water. Used in living things to form lipoproteins which are the main component of the walls, the membranes, of living cells.
LITHOSPHERE: Literally, sphere of stone; the solid crust of a planet below which is the molten magma of the interior.

METABOLISM: The process of catalysis by enzymes of chemical reactions within cells, the sum total of which is the metabolism of the complete organism.
MICROFOSSIL: A fossil of early microorganisms.
MICROORGANISM: A living thing consisting of a single cell, such as protozoa, yeast, algae, and bacteria.

NOBLE GAS: A member of the family of rare gases of the elements of group O of the periodic system of chemical elements; helium, neon, argon, krypton, xenon, and radon which do not readily tend to combine with other elements.
NUCLEOTIDE: A compound of a base derived from such as purine or pyridine, a pentose such as ribose, and phosphoric acid.

OBLIQUITY: Angular inclination of the equatorial plane of a planet to the plane of its orbit.
OCCULTATION: Obscuration from an observer's viewpoint of a distant body by a closer body passing in front of it; for example, when Earth's Moon passes in front of a distant star.
ORGANIC MOLECULE: A molecule containing carbon.
OUTFLOW CHANNEL: A channel which appears to have been formed by the flow of volatile material from a source such as an area of chaotic terrain or a breached crater wall.
OUTGASSING: The escape of gas from the material of a planet into its atmosphere, usually as the result of internal heating of the planet.

PERIAPSIS: The closest point to a planet or a satellite in a noncircular orbit around that planet or satellite.
PERIHELION: The closest point to the Sun in a noncircular orbit around the Sun.
PERMAFROST: A region in which water is permanently in the form of ice within the surface materials of a planet.
PHOTOAUTOTROPHY: Ability of a living organism to use light to assimilate carbon dioxide.
PHOTOSYNTHESIS: The process by which plants use solar energy to synthesize their structure from water and carbon dioxide.
PHYLA: A group classification for terrestrial life forms; e.g., all terrestrial animals fit into about a score of phyla.
PLANETESIMAL: A body made up of condensates from a primeval nebula. It is believed that the accretion of many of these bodies of various sizes produced the planets and their satellites about 4.5 billion years ago.
PLATE TECTONICS: The movement of plates of Earth's crust from spreading centers through which volcanic materials rise to the surface and produce more crust, thrusting the existing crustal plates apart.
PLAYA: A flat plain usually formed within an undrained desert basin by ephemeral lakes.
PLUVIAL PERIOD: A period of rain during which water is precipitated from the atmosphere onto a planetary surface.
PREBIOTIC MOLECULE: A carbon-rich molecule from which a living microorganism may have formed.
PRECAMBRIAN: A geologic period earlier than some 600 million years ago preceding the Cambrian period from the rocks of which we have evidence in the form of fossils of the first abundant marine invertebrates. Fossils of early terrestrial life forms from the precambrian period were discovered only comparatively recently.
PRECESSION: Change in the direction of the axis of rotation of a planet.
PYROLYTIC RELEASE EXPERIMENT: An experiment to seek evidence of microorganisms creating organic compounds from radioactively labeled carbon dioxide and carbon monoxide, with or without exposure to sunlight. After incubation, a high temperature was used to drive off gases for analysis. Later the soil sample was pyrolized to reveal any organic products remaining in the sample.

RADIOGENIC HEATING: Heating of the interior of a planet or satellite by the decay of naturally occurring radioactive elements within the planet or satellite.
RAMPART CRATER: Another name for a base surge crater, particularly one in which the flows appear to have developed ramparts at their extremities.
REGOLITH: The surface material of a planet, particularly one that has been mixed by the impact of bodies from space onto that surface.
RIFT VALLEY: A valley formed between fractures of a planet's crust.
RUNAWAY GREENHOUSE: A state in which the greenhouse effect is not limiting until the environment is changed in a catastrophic manner by the greenhouse effect; for example, until all the oceans on a planet have been lost as a result of the greenhouse warming.

SECONDARY CRATER: A crater formed by ejecta from another crater.
SHIELD VOLCANO: A volcanic mountain formed by the nonexplosive flow of very fluid lava. Examples are the terrestrial Hawaiian Islands and the Tharsis volcanoes of Mars.
SNC METEORITE: An unusual type of meteorite of a composition which suggests that it may have originated on another planet, particularly Mars.
STROMATOLITE: A rock built up layer by layer by microorganisms attracting and binding mineral grains present in shallow bodies of water.
SUBLIMATION: A change in state from a solid to a gas without passing through the liquid state; for example, when frozen carbon dioxide heats toward room temperature under normal atmospheric pressure on Earth.

TECTONICS: The molding of a planetary surface by forces acting from within the planet.
TELEOPERATION: Operation at a distance; the remote operation of a machine by commands transmitted to it by radio, electrical, or light signals.

WATER SAPPING: Emergence of water from below the surface with the formation of channels by that water.
WEATHERING: Changes to the surface of a planetary body by various processes such as wind erosion, water erosion, chemical actions, incoming radiation, and energetic particles.

ZEOLITE: Hydrated silicate; important in the process of terrestrial soils being able to retain the potassium which is essential for plant life.

INDEX